American Ideal

American Ideal

How *American Idol* Constructs Celebrity, Collective Identity, and American Discourses

Amanda Scheiner McClain

LEXINGTON BOOKS
Lanham • Boulder • New York • Toronto • Plymouth, UK

Published by Lexington Books
A wholly owned subsidiary of The Rowman & Littlefield Publishing Group, Inc.
4501 Forbes Boulevard, Suite 200, Lanham, Maryland 20706
www.lexingtonbooks.com

Estover Road, Plymouth PL6 7PY, United Kingdom

British Library Cataloguing in Publication Information Available

Library of Congress Cataloging-in-Publication Data

McClain, Amanda Scheiner, 1979–
American ideal : how American idol constructs celebrity, collective identity, and American discourses / by Amanda Scheiner McClain.
p. cm.
Includes bibliographical references.
ISBN 978-0-7391-6433-4 (cloth : alk. paper)—978-0-7391-6435-8 (electronic)
1. American idol (Television program) 2. Reality television programs—Social aspects—United States. 3. Fame—Social aspects—United States—History—21st century. 4. Television viewers—Social aspects—United States. I. Title.
PN1992.77.A56M33 2011
791.45'72—dc23 2011030756

∞™ The paper used in this publication meets the minimum requirements of American National Standard for Information Sciences—Permanence of Paper for Printed Library Materials, ANSI/NISO Z39.48-1992.

Printed in the United States of America

Contents

Acknowledgments

I would like to thank my dissertation advisor, Dr. Carolyn Kitch at Temple University, for her ceaseless enthusiasm for this project, editing skills, and never-ending encouragement. I'd also like to thank my dissertation committee members, Dr. Andy Mendelson and Dr. Nancy Morris, for helping to revise a dissertation that became very long. Thanks also to Adam Spector for his help in finding the *American Idol* seasons online. To my mother, Anne Scheiner: thanks for your interest in this research and the countless enjoyable next-day *American Idol* dissections. To my father, Dave Scheiner, and sister, Melissa Scheiner: thanks for your continuing encouragement and support. Lastly, thank you to my wonderful husband, Jordan McClain. I greatly appreciate your companionship and willingness to listen, to discuss, and to watch an endless amount of reality TV.

Chapter 1

American Idol

American Idol debuted in the summer of 2002 on the FOX network as just another show that could have been cancelled at any moment, unless it acquired the right audience. It did. By the end of the first season, the finale attracted more than 23 million viewers, the highest ratings FOX had received for a non-sports show in ten years (Bignell, 2005). The ratings continued to grow over the subsequent seasons. The third season finale, in 2004, drew 34 million viewers, earning the rank of the third-highest-rated program of the year, after the Super Bowl and the season finale of reality show *Joe Millionaire* (Bignell, 2005). The *American Idol* ratings dominance continued in 2005, with the fourth season averaging an audience of 26.8 million viewers over the course of the season, a 5 percent increase from the previous season. The 2006 fifth-season premiere attracted a sizeable audience of 35.5 million viewers (Keveney, 2006) and an average audience over the season of more than 30 million viewers (Garrity and Waddell, 2007). The fifth, sixth, and seventh seasons averaged more than 30 million viewers per night (Rosen, 2007; Serpe, 2007). In 2007, the sixth season of *American Idol* ranked as the most-watched TV series for the fourth year in a row (Serpe, 2007). Moreover, by the end of the 2007–2008 television season, the FOX network had the most viewers of any channel for the entire season, including the highly coveted 18-to-49-year-old viewer age bracket (de Moraes, 2008). *American Idol* was also the most time-shifted show, recorded on digital video recorders or TiVo and watched within seven days (Serjeant, 2008). One reason for the popularity of *American Idol* is its family appeal: it "lies at the intersection of youth and adult tastes, allowing everyone to show some expertise" (Jenkins, 2006, p. 81).

The *American Idol* ratings and popularity warrant study of the program and the communities engendered by it. As it dominates the television ratings, the contestants and judges appear in major news outlets, such as *People*, *Time*, *USA Today*, *The Today Show*, and *The Tonight Show*. Stories associated with *American Idol* are prevalent across all forms of media. *American Idol*, originally British and titled *Pop Idol*, has entered the national lexicon. At least partially due to its enormous success, the format of *Pop Idol* is replicated across the globe. The *American Idol* brand is conservatively valued at 2.5 billion dollars (Atkinson, 2007). Each season, Americans across the country debate the merits of contestants, at home, at work, and live on TV. *American Idol's* popularity is almost unavoidable. *American Idol* illustrates national ideas about American identity and celebrity and connotes prevalent social themes of 2000s American culture.

American Idol is a model of the American obsession with fame, due to the thousands of people who audition and the annual millions of viewers. Simultaneously, the apparent transparency of *American Idol* permits the examination of the preoccupation with celebrity through the visible formation of a nobody into a star. Previously, icons were created behind the scenes, requiring the right mixture of a carefully groomed image, sex appeal, and marketing to produce fan frenzy. *American Idol*, while mimicking the tried-and-true formula for star creation, attempts to reveal some of the operations behind the curtain by allowing audiences to seemingly view and aid in the idol creation process, while at the same time ironically utilizing the same marketing techniques to help create the ensuing popularity. Celebrity is created through appearances on the show, in other media, and through gaining viewers' adulation.

American Idol is a deceptively simple reality TV program. Each year thousands of people audition in six or seven cities across the country. TV viewers see presumably producer-selected auditioners sing for judges, who have included Paula Abdul, Randy Jackson, Simon Cowell, and an occasional guest judge. Kara DioGuardi joined the judges in season eight and Ellen DeGeneres replaced Paula Abdul in season nine. Season ten featured a revised judging panel consisting of Steven Tyler, Jennifer Lopez, and Randy Jackson. The panel of experts then selects about one hundred contestants to participate in "Hollywood week," which usually airs over two weeks. In Hollywood, the contestants perform individually and in groups. The judges eliminate contestants throughout the week, ultimately ending with a semifinal group of sixteen to twenty-four people, depending on the season.

Next, during the semifinal round, the contestants perform over the span of three weeks in front of a small studio audience; the judges continue to critique contestants' performances. However, they do not decide who is eliminated. From this point on, after "Hollywood week," the voting audience determines who will remain in the competition. Every week, each contestant is assigned a toll-free telephone number. Viewers may vote by calling or AT&T subscribers may vote via text messaging. In 2011, viewers were also able to vote online using their Facebook login. Typically, contestants sing on a two-hour Tuesday evening show and then people at home may vote for two hours following the show. The next night, Wednesday, the contestant who received the lowest total of votes is eliminated. Seasons one, two, and three also included a semi-final wildcard week, in which previously eliminated contestants were invited back to sing one more night in hopes of making the top twelve. The first season's top twelve contestants included one wildcard contestant chosen by the judges, while the next two seasons' top twelve finalists included four wildcard picks each, with one determined by each judge and one chosen by the audience. The wildcard round notably spared contestants Clay Aiken and Jennifer Hudson from early dismissal. Finally, the top twelve contestants perform weekly and audience members vote for their favorites, with the contestant receiving the lowest number of votes eliminated each week. This continues until the audience ultimately chooses the winner, who is revealed in a star-studded finale extravaganza.

Viewers may feel enfranchised through voting, but are ultimately subject to the same publicity machinations as audiences of the past. The entire program is promoted the same way the ultimate winner is—through a nationwide media campaign. Audience members may have agency in choosing the winner—however, they are still influenced by the presented narratives. The narratives are manipulated through editing, quotation choices, and stylistic decisions.

American Idol, while a continuation of the success of reality television, ushered in the new interactive era of the early 2000s. *American Idol* encourages audience participation, urging viewers to vote for their favorite performer, as well as engaging viewers through contests and online message boards. This participation engenders a distinctive type of audience member, one who seemingly has agency in the show's production. Voting for singers is a parallel of the American democratic election system. Audience agency is illustrated by the multitude of votes received. More than 70 million text-message votes were received in a two-hour period for the 2007 winner, Jordin Sparks (*American Idol*, May 23, 2007). Upon winning, Sparks sang a song written by an audience member and chosen by audience votes in an online contest. In 2008, audience members voted a combined 97.5 million times for David Archuleta and David Cook (de Moraes, 2008). Champion David Cook sang the winning song of the second *American Idol Songwriter* contest (Gundersen, 2008). A plethora of other shows also use audience interaction components to attract and maintain audiences.

AMERICAN IDOL: ECONOMICS

The program *American Idol* is a version of a British show, *Pop Idol. Pop Idol*, created by Simon Fuller, debuted in June of 2001 (Lisotta, 2006). The concept of the show echoes an earlier radio and TV show, *Amateur Hour*. The prevailing difference between *American Idol* and antecedent programs such as *Amateur Hour* and *Star Search* is continuity. As Stahl noted (2004, p. 212), *American Idol* combines "the democratic principle of an audience vote and the narrative principle of the month-long competition." The same contestants are featured over an entire television season, permitting home viewers an extended amount of time to become acquainted with the performers. Familiarity is engendered through interviews and packaged montages of their home lives. Additionally, the audience voting mechanism alleviates competition between contestants, fostering a familial group environment, akin to a drama series (Stahl, 2004).

The advertisements and ancillary products, as well as the records put out by *American Idol* artists, earn millions of dollars. By 2007, there were 10 million unique pieces of *American Idol* merchandise available for sale (Lieberman, 2007). In 2005, average prices for commercial air time during the show were $660,000 and $705,000 for a 30-second spot, the highest ad rates on network TV (Serwer and Hajim, 2005). By 2010, *American Idol* was still the most expensive program in which to buy commercial time (Steinberg, 2010). According to *Advertising Age*, in 2007 *American Idol* was

> conservatively valued at $2.5 billion as a franchise, [and] the *American Idol* empire already brings in $500 million a year in TV ad dollars, including a number of $30 million to $50 million core sponsorship packages, music sales, live tours-read: more sponsorship revenue as well as ticket sales-and an explosion of products from 40 licensees. (Atkinson, 2007, p. 1)

AMERICAN IDOL: OWNERSHIP

FremantleMedia, a part owner of *American Idol*, is one of the globe's largest TV and radio production companies and is a branch of the RTL Group, Europe's biggest media company. RTL Group is majority owned by Bertelsmann (FremantleMedia, April 16, 2007). Bertelsmann is one of the largest media conglomerates in the world; its holdings include several publishing houses and smaller media companies (Bagdikian, 2004; Resources: Who owns what, March 7, 2008). As a press release touted,

> FremantleMedia North America (FMNA) is the U.S. production division of global media giant FremantleMedia. Based in Burbank, California, FMNA produces entertaining and innovative television programs for network, cable and syndicated platforms, including the Emmy-nominated musical/reality phenomenon *American Idol* (FOX), *The Next Great American Band* (FOX), *America's Got Talent* (NBC), *American Inventor* (ABC), *Thank God You're Here* (NBC), *The Janice Dickinson Modeling Agency* (Oxygen), *Property Ladder* (TLC), *Family Feud* (syndicated), *Temptation* (syndicated), *Million Dollar Password* (CBS), and the longest-running game show in television history, *The Price Is Right* (CBS). (*American Idol* is going to Disney World! February 7, 2008).

FremantleMedia also produces and owns format rights to popular TV shows such as *Neighbours* and *The X-Factor*, as well as several others shown in countries around the globe (FremantleMedia, April 16, 2007). FremantleMedia owns the *American Idol* website, which is hosted, maintained, and marketed by FOX. It received more than 40 million unique visitors over the course of the 2006 season (Atkinson, 2007).

Another portion of *American Idol* is owned by 19 Entertainment. In 2005, 19 Entertainment was purchased by CKX, a media corporation previously known as SFX and associated with radio, concert venues, and concert production. CKX paid Simon Fuller $196 million dollars for 19 Entertainment, which also owns the reality game show *So You Think You Can Dance?* and manages the careers of David and Victoria Beckham (Serwer and Hajim, 2005). 19 Entertainment, primarily a talent management company, continues to run the daily production of *American Idol*. Simon Fuller was under contract to head 19 Entertainment until the year 2010. According to a press release announcing the creation of an *American Idol* themed attraction at Disney World,

> Simon Fuller's 19 Entertainment (of CKX, Inc., a publicly traded company listed on the NASDAQ National Market® under the ticker symbol 'CKXE') is recognized as a leading creator, provider and promoter of globally successful music-based entertainment, or 'entertainment brands.' The company has attracted a unique collection of expertise in people who work together to integrate and leverage activity across television, film, touring, merchandising, music publishing, recording, artist/writer and producer management, sponsorship and promotion. Simon Fuller is the most successful British music manager of all time having sold over 116 million CDs in North America alone, and is known as the creator of *American Idol*, the most valuable TV format in the world with an estimated value in excess of $2.5 billion. (*American Idol* is going to Disney World! February 7, 2008).

In 2004, *American Idol* and its associated products earned 19 Entertainment at least 30 million dollars (Serwer and Hajim, 2005). In 2010, Fuller and Barclays Capital banker Roger Jenkins bid $600 million in an unsuccessful attempt to buy CKX (Fuller bids for CKX takeover, 2010). Also in 2010, Fuller created a new management company, XIX Management, which currently acts as a consultant for *American Idol.* While 19 Entertainment, under the corporate umbrella of CKX, continues to operate the show and own a portion of it, XIX Management and Simon Fuller are managing the post-2010 *American Idol* victors' careers (Fuller to manage *American Idol* winner, 2011).

Furthermore, according to Nielsen SoundScan, winners Kelly Clarkson, Ruben Studdard, Fantasia Barrino, Carrie Underwood, and Taylor Hicks, as well as season-two runner-up Clay Aiken, have combined sales of more than 23 million copies of albums (Pinsker, 2007). They were all under contract to Sony BMG record labels. Sony BMG holds a roster of music labels, including Columbia, Epic, RCA, Arista, and J Records, among others (Sony BMG, April 16, 2007). The *American Idol* artist is assigned to the label most appropriate for his or her style of music. In January 2008, J Records dropped Ruben Studdard and Taylor Hicks. Studdard, at least, remains under 19 Entertainment management (Halperin, 2008). Sony BMG is an equal partnership between Sony and Bertelsmann (Sony BMG, April 16, 2007). Upon the completion of the program, Sony BMG, XIX Management, and 19 Entertainment have an exclusive agreement for right of first refusal for all recording and management contracts for participants of *American Idol.* XIX Management is under contract to manage the 2011 *American Idol* champion as well as future season winners. XIX Management and 19 Entertainment receive 10 percent of all profits as management of the artists (Serwer and Hajim, 2005). The careers of the recording artists are then promoted on the program. In theory, the success of the show propels the winners on a successful course of popularity, accomplishment, and fame.

In sum, *American Idol* is a program essentially owned by Bertelsmann and CKX, two very large media companies, with Bertelsmann being one of the largest media conglomerates in the world. It is broadcast in America through a synergistic agreement with another of the largest conglomerates in the world, NewsCorp. Additionally, *American Idol* partners with other consolidated media corporations. In an example of synergistic partnerships, in a cross-promotional segment, the 2007 *American Idol* contestants were shown a preview of the *Shrek 3* movie, learned about animation, and met *Shrek* star Antonio Banderas and Jeffery Katzenberg, CEO of DreamWorks. DreamWorks is owned by Paramount Universal Motion Picture Group, which in turn is owned by Viacom, another extremely large media conglomerate (Thottam, 2006). Jeffrey Katzenberg is the former head of another media conglomerate, the Disney Company. Not only does *American Idol* function as a vehicle to promote media owned by other conglomerates, but vertical and horizontal integration through its parent company are visible as well. Vertical integration of *American Idol* and Bertelsmann is evident by means of the inception, production, and eventual distribution of the show through various media subsidiaries in local formats throughout the world. Horizontal integration is evident through the recording contracts with Sony BMG, again earning money for the parent company through a different medium. The show earns revenue through a variety of sources, including product placement in the program, commercial advertising, franchises in diverse countries, licensed products, and through the synergies between record labels and the parent company. It is a symbol of the interactive convergence of media consolidation, with many channels of revenue all eventually streaming back to the parent company.

GLOBAL IDOL

American Idol is an originally British program that has been replicated around the globe. Although television has traditionally been exported and imported (Moran, 1998), the recent trend of exporting reality formats has taken hold over the past decade (Waisbord, 2007).

According to Waisbord (2007, p. 375) the popularity of reality television formats emerged for two reasons: the globalization of the television business model and the "efforts of international and domestic companies to deal with the resilience of national cultures." Globalized business dealings have increased the amount and intensity of connections between national television industries through transnational media companies (Waisbord, 2007). As media companies grow, the ease of selling the same format to multiple countries grows in tandem. Additionally, the privatization and deregulation of national television systems has contributed to the increase of imported formats (Dahlgren, 2000). Concurrently, in many countries the number of channels has dramatically increased, amplifying competition between media companies (Barker, 1997). While the spread of formats across the world may be considered evidence of globalization, the national character of each program speaks to the strength of regional cultures. Thus, television is both global and local (Waisbord, 2007).

In particular, European media companies have exported reality TV and quiz-show formats successfully. *Who Wants to be a Millionaire?* has been sold to 79 countries by British Pearson Media (Schneider, 2000). This is in contrast to the previous domination of the global television industry by American media firms (Waisbord, 2007). After the *Big Brother* format was successful in Holland, John de Mol Productions exported it to the German,

Portuguese, Spanish, Belgian, British, Swiss, Argentinian, Australian, and American markets, among others (Hill, 2007).

National adaptations of formats are popular for a variety of reasons. Some countries' restrictions on imported media favor locally produced shows, but allow for an imported format (Waisbord, 2007). Increasingly, once a program finds economic success in one country, producers replicate it in order to replicate the success (Holmes and Jermyn, 2004; Waisbord, 2007). Reality formats offer the enticement of adding audience members through an interactive website or voting. Additionally, money can be made from the newly minted celebrities, through magazine coverage and other possible deals (Waisbord, 2007).

Pop Idol, the progenitor of *American Idol*, is another good example of successful global replication. FremantleMedia, a distribution company, owns the distribution and format rights to *Pop Idol* as well as several other dramatic programs and game shows, and has worked to create local national versions of *Pop Idol*, including *American Idol*. In some countries FremantleMedia subsidiaries produce the show; in others, franchise-holders produce their own rendering (Guider, as cited in Raphael, 2004). There are more than 100 versions of *Pop Idol* throughout the world (Serwer and Hajim, 2005). The template of *Pop Idol* can be imported and adapted by any country, with the added benefit of already being a proven success, thus lowering the financial risk. Also, in many countries local versions of a program qualify to fulfill regulatory quotas for indigenous programming (Madger, 2004). Furthermore, the extensive potential for product placement encourages advertisers to invest in the program (Madger, 2004). *American Idol* features prominent product placement by Coca-Cola and AT & T. In another example, the *American Idol* contestants sing in Ford "music video" commercials throughout every season.

NATIONAL PROGRAMS

The globalization of television formats is a reflection of the global capitalist economy. As such, the national identities carried out through the programs are extensions of the worldwide economy (Waisbord, 2007). The programs are not nationalistic out of a patriotic duty, but rather "result from the intention to maximize profits while 'the national' continues to articulate cultural identities" (Waisbord, 2007, p. 382). Holmes and Jermyn (2004) asserted that the importation of reality-TV formats, which frequently originate in European countries, conflicts with the traditional notion of media imperialism.

Cultural identities may be shaped and reflected through television, but television producers aim to earn money, not construct identity. While "television programming recreates and perpetuates national sentiments" it truly is a manifestation of global economics and a testament to the tenacity of national ideals (Waisbord, 2007, p. 381). Waisbord (2007) also contends culturally relevant shows are more popular, perhaps due to the lack of cultural barriers.

Moreover, Waisbord (2007) alleges that reality-TV and game-show formats are designed to transcend national boundaries—they embody neutral values. Following Billig's (1995) concept of banal nationalism, the very language of the national production helps advocate national ideas (Waisbord, 2007). While generally unthought-of, language is the basis of any program. Other aspects of programs add to the nationalistic ideas endorsed.

RESEARCH QUESTIONS

The research questions posed in this book are:

RQ 1: What contestant narratives are present in *American Idol*? Within these narratives, what do ideals and discourses imply about America?

RQ 2: In *American Idol*, how is celebrity created and represented?

RQ 3: How may *American Idol* illustrate American collective identity? What do the online discussants' conversations and contestant narratives connote about possible collective identity of an *American Idol* audience?

This book is organized topically. The next chapter, two, is an overview of theory and methods. Chapter three is an assessment of the program's American rags-to-riches journey, while chapter four is a catalog of the show's archetypes. Chapters five and six are an evaluation of *American Idol* celebrity; the former is an investigation of *American Idol's* authenticity discourses, while the latter is an exploration of *American Idol's* construction of celebrity confirmation. Next, chapter seven is a study of collective identity as discerned within the show discourse. Chapter eight continues this theme through an analysis of collective identity present on the *American Idol* message boards. The final aspect of this research is a press coverage discourse analysis explicated in chapter nine. Chapter ten is a summary of findings and includes the conclusion.

I believe this research into the popular artifact of *American Idol* is important in order to investigate American culture. Popular television reveals constructed American expectations, ideals, and hopes for the future, no matter how close or far these may be from reality. Through this analysis of *American Idol*, ideas about contemporary notions of celebrity, American collective identity, and American ideals are ascertained. The monstrous popularity of *American Idol* demands study of the program and the ideals contained within it.

Chapter 2

Overview of Theory and Methods

Television is a pervasive element in American life, present in almost every home. Millions of people watch hours of television daily. The integral nature of television in everyday life calls for a study of how it may affect Americans. Indeed, a multitude of television studies exist. Newcomb (2007) noted that television studies are a burgeoning field of research, with many scholars studying the production, reception, and ideological apparatus present within it. It is important to study television because "Television-as-culture is a crucial part of the social dynamics by which the social structure maintains itself in a constant process of production and reproduction: meanings, popular pleasures, and their circulation are therefore part and parcel of this social structure" (Fiske, 1987, p. 1). Television helps shape, reflect, and maintain culture. Therefore, studying television may reveal insight into aspects of culture. This study examines what messages may lie within television, particularly *American Idol.*

Television is a "cultural agent," a "provoker and circulator of meanings" (Fiske, 1987, p. 1). Moreover, Fiske (2000, p. 220) made the large claim that television attempts to "make meanings that serve the dominant interests in society." Thus, meanings reinforce the existent social structure, firmly establishing the status quo as "unchallengeable and unchangeable" (Fiske, 1987, p. 36). Meanings are established through common cultural codes: "A code is a rule-governed system of signs, whose rules and conventions are shared amongst members of a culture, and which is used to generate and circulate meanings in and for that culture" (Fiske, 2000, p. 221). Being a part of a culture, living in it, and growing up within it, engenders common understanding among members. Members of the same culture can watch a television program and extract the dominant meanings based on their shared culture. Accordingly, people from different cultures may understand meanings disparately, as Ang (1985) demonstrated.

While dominant social meanings are typically produced and repeated on television, it is a polysemic medium (Fiske, 1987). However, "the text delineates the terrain within which meanings may be made and proffers some meanings more vigorously than others" (Fiske, 1987, p. 16). The preferred set of meanings resonates with dominant ideology, although audience members' place in social structure may prompt alternate meanings.

The economics of television frequently dictate the content (Fiske, 1987). As different social groups may detect varied meanings, producers tend to make television content broad, in order to appeal to as large an audience as possible. Television employs widely recognized conventions, which appeal to viewers through familiarity (Fiske, 1987). The blander the content, the wider the potential audience base. The larger the audience, the more producers can charge for advertising. The lowest common cultural denominator is usually embraced in order to appeal to a large audience.

As content needs to be broadly understood by the mass audience, producers rely on familiarity. Subsequently, popular cultural standards mandate the codes present in television. Conventional social codes are directly related to ideological codes (Fiske, 2000). Fiske uses the example of a theft on a television program. Generally, the audience understands theft as a crime of personal property. Thus, theft directly corresponds to the ideology of capitalism, in that property is understood as being owned by a person (Fiske, 2000).

Codes of meaning, while visible in all aspects of television shows, are particularly present in casting. As Fiske asserted, "characters on television are not just representations of individual people but are encodings of ideology 'embodiments of ideological values'" (Fiske, 1987, p. 9). Qualities are ascertained through codes, conventionally and intertextually. Just as there is a meaningful code associated with stereotypes, other accepted and ideological codes are present within other aspects of television programs. The "ideological codes work to organize the other codes into reproducing a congruent and coherent set of meanings that constitute the common sense of a society" (Fiske, 2000, p. 223). Thus, television and reality work together to create the common culture, cyclically affecting each other. Television is often viewed as both a source and an effect of social and cultural adjustments (Newcomb, 2007).

Other aspects help induce audience understanding as well. Settings and costumes help compose characters (Fiske, 2000). Costumes provide a means of identification for the audience. They can discern who the hero is and identify with him based on costume. Setting and dress can manifest ideas of class, morality, and attractiveness. Concepts of race and gender are also built into physical codes. Thus, "abstract ideological codes are condensed into a set of material social ones" (Fiske, 2000, p. 227).

Codes are often expressed through narratives—television is primarily a narrative medium (Fiske, 1987). Narratives are chronological and easy for audiences to understand. They work as a sense-making mechanism for audiences (Fiske, 1987). Through a narrative format, events link together and form meaningful relationships. Importantly, narratives are an essential component of myths, which naturalize aspects of reality. Myths explain societal structure and are present not only in television but also in other media and folklore (Barthes, 1957/1995). Fiske (1987, p. 135) contended that utilizing narrative analysis to decode myths "emphasize[s] the cultural-ideological system that underlies the syntagmatic flow of the narrative"

While narratives compose myths in all types of media, some particular narratives are unique to the television medium. The television series is open and usually incomplete, lacking the closure of a film or novel (Feuer, 1986). Similarly, the news also is a continual, never ending narrative (Fiske, 1987). Television narratives may mimic off-screen conventions, such as placing more importance on masculine ideals (Feuer, 1986). Furthermore, most conflicts are set against the unstated, implicitly understood happiness of the conventional nuclear family (Feuer, 1986). Narratives and myths, which are further explained in the methods discussion, are both means that *American Idol* uses to connote meaning.

Television also stresses the occurrence of real time. Television typically is set in the same time frame as off-screen reality. Movies may compress years into two hours—television generally has much less time compression. This makes television more applicable to contemporary life than film or novels (Fiske, 1987). Similarly, television characters have a past, present, and future. Audiences will see them next week, or next season. Television characters are easier to identify with than film characters, whose lives cease at the rise of the credits (Fiske, 1987). *American Idol* operates in real time. When a week passes in off-screen time, it does for the contestants as well.

REALITY TELEVISION

American Idol is an example of reality TV television programming. Since the early 1990s this genre has dramatically risen in popularity. Historic precedents of reality TV include *Amateur Hour*, *Candid Camera*, *Star Search*, *That's Incredible,* and the many game shows, dating shows, and make-over programs from yesteryear. Stahl (2004, p. 214) noted that *American Idol* differs from its game-show predecessors through its "protracted, intimate, reality TV focus on the subjectivities and biographies of its contestants, their private relationships, and the rigors they were put through during the elimination process" The genre of reality TV, and subsequently *American Idol*, is focused on "real" people and the intimate minutiae of their lives. Therefore, reality TV is often considered an amalgamation of voyeurism and entertainment (Andrejevic, 2004; Dauncey, 1996; Dovey, 2000; Ellis, 1992). Murray and Ouellette, accordingly, defined reality television as the "fusion of popular entertainment with a self-conscious claim to the discourse of the real" (2004, p. 2). Reality TV bridges the staged event and actuality through mediated intimacy (Kavka, 2005). The reality programs feature "ordinary people doing ordinary things" creating "the people show" (Tincknell and Raghuram, 2004, p. 258).

Often a game-show aspect is present in reality TV. This game-show segment of the reality-TV genre is illustrated through programs like *Survivor*, *Big Brother*, *The Apprentice*, *The Amazing Race*, and dating shows such as *The Bachelor*. Another division of the genre is makeover shows, such as *Extreme Makeover: Home Edition, Queer Eye for the Straight Guy*, *What Not to Wear*, and *The Swan*. Still another portion of the genre is the docusoap (a combination of documentary and soap opera), evident in *The Real World, High School Reunion,* and *Sorority Life* (Murray and Ouellette, 2004). Other docusoaps are *Snoop Dogg's Father Hood, Keeping Up with the Kardashians*, *The Girls Next Door,* and *Kendra*. The subgenres of reality TV are frequently debated, as shows often defy categorization due to the intertextual self-reflexivity common among such programs (Coles, 2000). For example, *The Amazing Race* is a game show where contestants race to win, yet also permits a seemingly uninhibited glimpse into contestants' lives, akin to a docusoap. Contestants may emulate contestants from other shows; they understand what is expected of them.

HISTORY OF REALITY TV

Some authors, such as Kavka and West (2004), argued that reality TV is ahistorical and takes place out of the context of chronological time. It does not need to relate to any off-screen context. Similarly, Nichols (1994) stated that reality TV takes place only in current time, lacking historical perspective. While reality TV may privilege the present over the past, it does have historical antecedents. Friedman (2002) placed the origin as the live radio program of the 1920s. Other authors (such as Fishman and Cavender, 1998; Leishman and Mason, 2002) place real-crime programs as primary ancestors of reality TV. Jermyn (2004, p. 72) posited that real-crime programming and reality TV share the "spectacle of actuality" Early examples of real-crime programming are the British *Crimewatch*, first airing in 1983, and the American *Unsolved Mysteries*, which debuted in 1987 (Jermyn, 2004). These programs largely featured dramatized reconstructions of events that actually occurred (Kilborn, 1994).

Cavender (2004) traced the crime drama influence to the 1940s radio and television police procedural programs, based on real cases, leading to *Dragnet* in the 1950s, which was also based on real crimes.

However, ordinary people, not actors or journalists, have long appeared in television "news, quiz shows, and documentary" genres (Holmes, 2004a, p. 113). Many vaudeville performers first appeared on live radio shows and then transitioned to TV, in programs such as *Texaco Star Theater* (Brooks and Marsh, 2003). *Texaco Star Theater* aired on radio from 1940 to 1948 and on television during the years 1948 to 1956. Other stage vaudeville shows reappeared on television in variety and game-show formats. Moreover, early radio and television game shows like *Queen for a Day* featured non-actors—working-class women and men. Indeed, the confessional narratives present in early game shows such as *Queen for a Day*, *Strike it Rich*, *It Could Be You*, and *The Big Payoff* (Cassidy, 2001) seem a clear historical lineage to today's tell-all, uninhibited reality-TV shows. People appeared on camera as themselves, relating their stories and why they should win. Many of these game shows began on the radio as well, moving to TV in the late 1940s and early 1950s.

In another example, Clissold (2004) credited the television program *Candid Camera* as a primary antecedent of today's reality television. *Candid Camera* first aired as the radio show *Candid Microphone* in 1948 and moved to television as *Candid Camera* in 1949. The show combined two Cold War fears: "surveillance-anxiety—the constant fear that one's actions were being monitored" and "simulation anxiety—the inability to distinguish between the real and not real in the age of technological reproduction" (Clissold, 2004, p. 33). Different versions of these discourses, not fears, can be found in contemporary reality television.

Clissold (2004) asserted that *Candid Camera* was the first television program to normalize surveillance, a theme Andrejevic (2004) elaborated upon in relation to reality TV. Through *Candid Camera*, audiences became accustomed to surreptitious recording, only revealed later to the participant. Thus, surveillance fears were alleviated through comedy, while still reminding audiences that they were constantly being watched (Clissold, 2004). While a clear ancestor of all reality TV, *Candid Camera* is directly related to programs like *TV's Bloopers and Practical Jokes,* which aired as a regular series in the 1980s and as specials throughout the 1990s and 2000s, and more recent programs such as *The Tom Green Show*, *Jackass*, *Punked*, and *Pranked.*

Candid Camera also was one of the first programs to turn the camera away from actors and focus on normal individuals. Average Americans, through signing the release form permitting *Candid Camera* to broadcast their scenes, became celebrated on film (Clissold, 2004). The release "effectively reversed the traditional power dynamic associated with hidden surveillance" (Clissold, 2004, p. 41). In effect, ownership of the recorded scenes and a choice of whether or not to allow the producers to air them mitigated the negative ideas associated with being secretly recorded. Through the agency of choice "helpless victimization [was transformed] into willing participation" (Clissold, 2004, p. 42). Thus, the surveillance power dynamic was transformed into a favorable one for the average citizen.

Reality TV enacts a similar transformation of power. Reality TV provides power not only to the writers or producers but also to the people who create the action, the participants who embrace being recorded. The stars of reality TV decide how to behave, creating plot, although within the constructed environment created by the show producers. Therefore, surveillance can engender a new power in the citizen stars, in effect democratizing television power relations. Of course, producers and editors hold ultimate control over participants' depictions.

The 1990s and 2000s saw the popularity of reality TV exponentially increase. During the 1990s, the docusoap began to earn high ratings (Dovey, 2000; Hill, 2005, 2007; Holmes and Jermyn, 2004). One of the fathers of current reality-TV programming and a first docusoap, *The Real World,* premiered in 1992. *The Real World* was actually envisioned as a soap opera; when the producers ran out of money, they elected to use real people instead of actors (Pullen, 2004). The docusoap programs of the 1990s and early 2000s typically followed a core group of people going about their daily lives. Docusoap shows of this era included *The Osbournes*, *The Anna Nicole Show*, *Laguna Beach*, *The Hills*, and various incarnations of *Big Brother* (Holmes and Jermyn, 2004). The docusoap, as evidenced by these programs, combines the "observational rhetoric of documentary . . . with the discourses of display and performance" (Holmes and Jermyn, 2004, p. 5). The one common aspect all programs claiming the label reality TV share is the "claim to 'the real'" (Holmes and Jermyn, 2004, p. 5).

There is a good deal of debate about the nature of reality TV and why it exists. The political-economic vantage point notes that reality TV emerged as a cost-cutting solution to audience fragmentation, the growing number of media channels, and broadcast television's corporate debt (Barnfield, 2002; Raphael, 2004). These and other related factors led to the low-cost production of reality TV, which requires significantly less overhead and generally grosses substantial revenue. Simultaneously, increased pressure on producers and increased competition due to the proliferation of channels also helped stimulate the increase in the amount of reality-TV programs (Dovey, 2000). Nonetheless, Creamer (2001) alleged that when costs of marketing and advertising are taken into consideration, reality TV may not be as cheap as expected. In addition, the television writers' strike of 2007–2008 helped add to the number of reality-television programs aired those and in subsequent years (Tucker, 2007).

Another means of examining this subject, particularly as related to documentary, is not as real versus constructed but rather as a blurring of the private and public spheres (Stephens, 2004). Aspects of personal lives that had previously been confined to the home, away from prying eyes, are now in the public discourse of television. Programs appearing on the TLC channel, such as *A Wedding Story* and *A Baby Story,* follow events, marriage and motherhood, that were typically considered feminine and located in the private sphere (Stephens, 2004). The airing of these personal milestones on television destroys the demarcation between private and public. Thus, what was previously private is now considered reality.

Documentaries generally have lower production costs than other types of feature films. Thus, many authors assert reality TV is an emulation of documentary film style for economic reasons (for example, Bignell, 2005; Kompare, 2004). Due to the documentary style of reality TV, fiction and non-fiction blur, resulting in a new form of media. The documentary characteristics present in reality TV may cue viewers to assume that scenes are real—but the nature of reality TV is that everything is constructed, except possibly emotions and talent. Bignell (2005) contended that the term reality TV denotes a cultural phenomenon of altering relationships between individuals and society, as encouraged by television. Reality TV is thus a category of identification for these types of programs visible across many TV schedules throughout the world. Along these lines, Friedman (2002) argued that reality television has been present since the inception of television, due to the live broadcasting typical of early TV.

Today, most reality TV is not broadcast live. As a consequence, the genre designation of reality TV is conceived as a marketing tool, not a shift in television production (Friedman, 2002). Nevertheless, despite the consideration of the label as a promotional device, it still describes an accepted category of TV programming, as indicated by the abundance of popular and academic media texts devoted to it.

REAL OR NOT?

Typically, shows categorized as reality programming must fit two requirements. The participants on the show must not adhere to a written script and must be willing to have their daily lives broadcast on television (Andrejevic, 2004). *American Idol* offers seemingly unscripted interviews and provides admittance to contestants' home lives through video packages, interviews, and personal details, usually images of family, home life, and often military service. The details of their home lives make the participants "real," as does their behavior in the interview segments. Andrejevic's (2004) study of *Big Brother*, a program in which the home audience voted participants off the show, indicated that the more "real" contestants were, the more likely they were to win. According to the audience study, the winner of *Big Brother* was the contestant the audience judged to be the most "authentic." Thus, genuineness and sincerity are valued, although the degree of authenticity in the presence of cameras is disputable. In fact, the constructed nature of the entire situation may belie the whole concept of realness (Friedman, 2002). All the same, the artificiality of the construction of the program may be considered countered through the use of non-actors (Andrejevic, 2002).

The non-actors often become celebrities. In the case of *American Idol*, the contestants strive to be recognized for their singing and consequently to become celebrities. The celebrity as a sign contains "tension between authentic and false cultural value" (Marshall, 1997, p. xi). The celebrity is authentic in existence and emotions, yet also constructed. Therefore the celebrity is both authentic and false. Similarly, the contestants are construed as real, no matter the constraints of the artificial situation (Biressi and Nunn, 2005).

In reality TV, "fact" and "fiction" are continuously being negotiated. What audiences understand to be real or not is continually being reconstructed (Holmes and Jermyn, 2004). Thus, "reality" is open to interpretation and is constantly adapting. As audiences became more familiar with reality-TV formats, the programs become increasingly more intertextual, creating a revisionary cycle of understanding. Understanding season seven of *American Idol* is more intuitive after seeing a previous season. Moreover, in-depth comprehension of programs like *So You Think You Can Dance* may be affected by previous viewing of *American Idol.* Shows intertextually reference other programs within their genre.

CELEBRITY

Contemporary American culture is rife with celebrities, emanating from fields such as politics and entertainment, and ranging from Barack Obama to Madonna. Modern-day fame rose with the popularity of mass media (Gamson, 1994). Fame and celebrity, in addition to the culture industry as a whole, were criticized early by cultural theorists Horkheimer and Adorno (1944/1972) as means of social obedience conceived through "mass deception," a theme on which many current authors elaborate. Fame is an important aspect of American culture—as Caldwell (2006, p. 50) noted, "the rickety foundations of celebrity and luck have today been accepted as replacements for the old frontier individualism and the Puritan work ethic that once so securely supported much of the weight of the full-blown American dream." Celebrity is one of American contemporary society's dreams, along with success and wealth. According to many definitions, this dream may be attainable through perseverance and hard work, but also requires one necessary ingredient—luck. Would-be entertainers may toil unnoticed for years, waiting for the lucky audition that allows achievement of the celebrity dream. *American Idol* capitalizes on the drive for renown, mining the yearly thousands of auditions for potential stars.

Fifty years ago, Boorstin (1961/1992, p. 57) defined a celebrity as a "person who is known for his well-knownness." His comments on celebrity still ring true today, underscoring the unchanging ideal of celebrity. He observed, "Celebrity is made by simple familiarity, induced and re-enforced by public means. The celebrity therefore is the perfect embodiment of tautology: the most familiar is the most familiar" (p. 61). In this sense, once a person appears famous, the more famous he or she can become. Moreover, fame is not dependent upon talent, skill, or any inherent quality (Boorstin, 1961/1992). Thus, being famous is simply being known—the reasons for fame are inconsequential. Similarly, Turner (2004) contended that celebrity is not any combination of attributes, but rather is created and sustained through discourse about the celebrity. Turner defined celebrity as "a genre of representation and a discursive effect" and "a cultural formation that has a social function we can better understand."

Historically, celebrities and famous people have always existed. The terms fame and celebrity are often synonymous and in this book are used interchangeably. Early Romans celebrated their heroes (dead and alive) by reproducing their likenesses on coins (Gamson, 1994). Braudy (1997) argued that Alexander the Great was the first celebrity. Alexander the Great, living circa 350 B.C., managed his image through early media forms such as statues, busts, and coins. Typically, early celebrities were politicians or royalty (Braudy, 1997; Burke, 1992; Evans, 2005b).

In his discussion of the historic causes of contemporary celebrity, Rojek (2001, p. 11) associated the decline of royalty with the rise of the "ideology of the common man." The rights of man, capitalism, and democracies replaced monarchies through various revolutions in the eighteenth and nineteenth centuries (Marshall, 1997). Celebrities replaced monarchical figures, symbolically supplanting kings with common men, thus ideologically supporting anti-monarchy political ideals. This "ideology legitimated the political system and sustained business and industry," guaranteeing celebrity commodification (Rojek, 2001, p. 13). The rise of celebrity is also linked to the increase of a money-based economy and the decrease in membership in organized religion (Rojek, 2001). Thus, celebrity is inextricably connected to both democracy and capitalism (Gamson, 1994; Marshall, 1997; Rojek, 2001). The rise to fame is akin to the democratic rise to power and may have similar outcomes in the holding or relinquishment of power; correspondingly, fame and celebrity buttress capitalism.

The rise of mass media created the current incarnation of celebrity, beginning with the popularization of the newspaper in the early 1830s (Gamson, 1994). During the 1860s, P.T. Barnum pioneered the art of publicity, creating celebrities out of circus performers (Gamson, 1994). In the 1890s, competition between newspapers led to sensationalism and the creation of celebrities in an effort to attract readers. By the early twentieth century, photography began to be used in media, adding a visual element (Gamson, 1994). The start of the twentieth century saw the rise of the film studio system, contemporary celebrity, and the modern publicity apparatus (de Cordova, 1990; Gamson, 1994). Stars were created, managed, and publicized by the film studios. Popular magazines helped form star images through interviews, stories, and pictures. The film studios, publicity departments, and fan magazines worked together to produce and circulate knowledge of actors, creating the initial twentieth-century mass-media stars (de Cordova, 1990; Gamson, 1994). Actresses and actors such as Charlie Chaplin, Mary Pickford, and Douglas Fairbanks are examples of early film celebrities, popular in the 1910s.

NARRATIVES OF FAME

Gamson (1994) used two common narratives to explain the achievement of fame. In the first narrative, merit earns fame, requiring the star to have an inherent greatness or talent. In the second narrative, publicity produces fame. The two narratives coexist within media texts (Gamson, 1994).

The first narrative discusses inherent star quality. Gamson (1994, p. 15) claimed that early in the twentieth century, the first modern celebrities were positioned as the "great and gifted." In this narrative, a person is a celebrity because he or she has the elusive, unnamed attribute that deserves fame. Alberoni (1962/1972) called this characteristic charisma. The popular press names it "star quality." No matter what it is called, the possessor of the special attribute is, and has always been, a star. Publicity may get the person noticed, but ultimately "fame is deserved or earned, related to achievement or quality" (Gamson, 1994, p. 15). This inherent quality is described as "unique, [and] ultimately inexplicable," a "wonderful gift of nature," and must lead to fame

(Rojek, 2001, p. 29–30). The narrative of merited fame asserts that "greatness is built in, it is who you are" (Gamson, 1994, p. 32). Hard work by the celebrity is not a part of this narrative. Once an unknown person receives her lucky break, her natural greatness is discovered and lauded. Her natural talent deserves celebrity (Gamson, 1994). Modern examples may include Bono, Bruce Springsteen, Venus Williams, and Michael Jordan. The narratives about all these celebrities include natural skills or talents that grant celebrity status. Beauty can also grant fame—for example, Brad Pitt and Gisele Bundchen may be considered famous due to sex appeal and handsomeness, as opposed to talent or luck.

Gamson (1994, p. 57–58) stated that in the second narrative, "fame is artificially producible and produced, well-knownness a salable and sold commodity, achievement divorceable and divorced from renown." Media professionals, transforming fame into a commodity and a star into an image, construct fame. Image, in the constructed-fame narrative, replaces quality in the merited-fame narrative (Gamson, 1994). No talent or inherent skill is necessary. For a celebrity, "what is developed and sold is the capacity to command attention," which in turn, may be considered fame itself (Gamson, 1994, p. 58).

Fame-as-commodity positions the star as a brand, therefore marketed as a new product (Dyer, 1986/2004; Gamson, 1994; Turner, 2004). Whether fame is achieved by merit or by media professionals, the image of the star must be manipulated to create the best possible aura for sales, creating a fictionalized version of the person (Gamson, 1994). Marshall (1997, p. 245) stated, "The celebrity operates as a brand name for the organization of production and consumption of cultural commodities." Once a star's image is established, it can be associated with a variety of products, transferring the celebrity image attributes to the product. Thus, the star/image is a brand unto itself and is expressed through a multitude of media texts (Dyer, 1986/2004; Gamson, 1994). Correspondingly, Biressi and Nunn (2005) posited that material and sartorial aspects of fame must be embraced to establish celebrity image, particularly if the aspiring star possesses a working-class background. The brand version of a star, validated through the accessories of fame, may be true for fame resulting from either of the two narratives, and no matter where or how celebrity emanates.

Achieving celebrity is inherently tied to an adoring public (Dyer, 1986/2004; Rojek, 2001). Audiences can be fickle; fame only lasts as long as the audience remains. To this extent fame is often mediated by "cultural intermediaries," a collective term for "agents, publicists, marketing personnel, promoters, photographers, fitness trainers, wardrobe staff, cosmetic experts, and personal assistants" (Rojek, 2001, p. 10–11). The task of the cultural intermediaries is to "concoct the public presentation of celebrity personalities that will result in an enduring appeal for the audience of fans" (Rojek, 2001, p. 10). Even a celebrity with inherent qualities that engender celebrity must maintain his or her fame through presenting an audience-friendly persona. The system of celebrity creation and maintenance is self-perpetuating and symbiotic. Gamson (1994) pointed out that publicists and agents need press for their clients, and similarly, the press requires the fodder of star interviews and actions. Cyclically, the more famous celebrities are, the more they appear in the press, and the more famous they become.

While much of a celebrity image is constructed through media presentation, it is also built through audience readings of dominant cultural representation (Dyer, 1986/2004; Gamson, 1994; Marshall, 1997). Image, therefore, is constructed through media *and* audience. The audience is responsible for accepting a celebrity as famous. Without the audience, celebrity deteriorates. Celebrities may strive to achieve audience acceptance by embodying cultural ideals. Thus, Rojek (2001) argued that past celebrities and current celebrities frequently have expressed two selves, their public persona and their inner "me," their true self. This acknowledges that the famous self is an image, constructed solely for reasons relating to fame. This constructed image is necessary to present an audience-acceptable persona. For example, closeted stars have often hid their sexuality, creating one self for public consumption and a second self for personal life.

Alternately, Dyer (1998) asserted that often film-star fame is based on a conflation of private and public life. Audiences may confuse on-screen characters' lives with off-screen actors' lives. Fame is often contingent on public narratives of private lives. This rings particularly true for today's reality-TV celebrities—they are famous for revealing their private lives, their seemingly true selves. Modern conceptions of fame are considered fleeting (Rojek, 2001) and sometimes hinge on the private life of a star rather than performative media work (Geraghty, 2000). Reality TV functions in a similar way, blurring any remaining boundaries between public and private selves, turning the private self into a public self, creating fame through personal exhibition.

American Idol invites the audience backstage to witness and contribute to the transformation of an average yet talented person into a star. There, Rojek's (2001) conception of a public persona/personal self blurs. The public persona and personal self amalgamate through the *American Idol* narrative of star creation. Likewise, Gamson's (1994) constructed-fame narrative and deserved-fame narrative also combine in *American Idol*. A contestant deserves to be a star due to inherent talent and innate star quality. Concurrently, the reality-TV component makes the personal aspects of the self common knowledge—the private persona is now public and celebrated. These discourses of authenticity (discussed in chapter 5) contribute to the ordinariness of the contestants, an integral part of becoming a reality-TV star.

Private doubts and fears join with the contestant's background, family, and religion to humanize the contestant. The ordinary becomes extraordinary. *American Idol* makes the star process seemingly transparent, asking the audience which talented contestant they think will make the best celebrity based on their consumer history of celebrity experience. The audiences are expert judges, confirming the celebrity of the winner. Through observing the contestant narratives and voting, transforming a nobody into a celebrity, the show asks the audience to partake in the process. The narrative of media creating celebrities is turned on its head. The audience and the producers of *American Idol* work together, complicit in the constructed confirmation of a star.

CELEBRITY AND TELEVISION

Many authors (for example, Ellis, 1992; Gamson, 1994; Holmes, 2004a; Marshall, 1997) have noted the differences between the conception of a film star and a television celebrity. Marshall (1997) also considers music celebrities to be a third category. Film stars are considered truly famous, while the medium of television denies its actors the same fame (Ellis, 1992; Langer, 1981). Music celebrities are considered famous based on the judgment of their authenticity, which varies according to style, time period, and music genre (Marshall, 1997).

Thus, the medium dictates the type of celebrity it produces. Music celebrity is created through technology of production and reception, commodification, and the audience's reception of the music itself (Marshall, 1997). Additionally, music is a more definable talent than acting. Music celebrities try to differentiate themselves from each other and the audience, while remaining committed to their relationship with the audience (Marshall, 1997). Television engenders familiarity and intimacy by entering into homes, while film creates the quality of enigmatic presence, allowing more of an awesome distance between the audience and the star (Holmes, 2004a; Langer, 1981; Marshall, 1997). This distance could be considered Walter Benjamin's (1936/1973) concept of aura, an air of singularity and distinctiveness found in one-of-kind, original, high-culture art. While film stars may be literally bigger than life on the screen, television performers are reduced in size, adding to their accessibility (Gamson, 1994). On television, individual characteristics are exaggerated through close-ups and dialogue, making the performer and audience interactions seem intimate (Langer, 1981). The familiarization process of television as a medium has several components: domestic family viewing, live or simulated live format, recognizable hosts, the task of selling the audience to advertisers which makes programs as inclusive as possible, serialization, and a focus on the personal (Marshall, 1997). Repetition adds to the familiarity (Gamson, 1994). Together, these attributes work to generate television celebrities. Celebrities who "emerge from television either service the process of familiarization (i.e., the hosts) or work in the domain of the familiar and the intimate (i.e., situation comedies or soap operas)" (Marshall, 1997, p. 131). Doubtlessly, reality television is an extension of this system as well. Reality TV literally focuses on average and consequently familiar people on TV.

Holmes (2004a) suggested the intensity of daily on-camera tasks in *Big Brother* and *Survivor* allows the audience intimate and banal knowledge of the contestant, fully removing any distance-created aura. This intimate type of fame is hinged upon the idea that the person on reality TV is interchangeable with the audience member. The "intimate relationship between participant and viewer" as well as the agency given to the audience through interactive components of the program, shape the form of celebrity caused by intense "familiarity, repetition and lack of 'aura'" (Holmes, 2004a, p. 117). The *American Idol* celebrities are a combination of music and television celebrities. They are seemingly average people, interchangeable with an audience member, but are distanced through music, a definable and difficult to achieve talent.

Reality TV's familiarity and interactivity foster a relationship that bestows fame upon ordinary people. This is a parallel of the frequent requirement that a celebrity be both ordinary and extraordinary (Holmes, 2004a). The star is special, but the audience must be able to identify with a star (Clarke, 1987; Dyer, 1998; Ellis, 1992). Dyer (1998) asserted that this ordinary/extraordinary duality is present in the merited-rise-to-fame narrative; the ordinary person receives a lucky break, becoming extraordinary. However, reality-TV participants must adhere to their ordinary attributes in order to maintain their claim to fame—that they are ordinary. *American Idol* contestants must be both talented and ordinary in order to achieve celebrity.

CELEBRITY AND AUTHENTICITY

The notion of authenticity is present in many conceptions of celebrity (Dyer, 1998; Ellis, 1992; Holmes, 2004a). Authenticity makes the star more than simply an image (Gamson, 1994). Realness gives the audience a way to identify with the star and a reason to like the star. Early film stars were promoted as being ordinary people, with average hobbies and pastimes, akin to their audiences (Gamson, 1994). In the 1930s, as audiences became more aware of the fabricated nature of image, publicity began to stress how authentically normal the celebrities were (Gamson, 1994). However, it still may be difficult to distinguish between an artificial image and authentic reality (Gamson, 1994).

In reality TV, there exists an emphasis on genuineness that belies the stars' constructed personas (Holmes, 2004a). Most people are aware of the constructed nature of reality TV—situations are entirely manufactured and do not occur in real life. Authenticity mediates this constructed nature and facilitates the construction of celebrity (Andrejevic, 2004). Despite audiences' awareness of the program's construction, they watch in hopes of seeing "moments of authenticity," or instances of truthful emotion (Hill, 2002, p. 337). Presumably, emotions are hard to fake and thus are important indicators of contestants' personalities. Audiences regard the constructed design of the program as inevitable and therefore of no great importance (Hill, 2007). Personality, indicated by emotions, is a component of who deserves to be a contestant or a winner. The manufactured image is created in two ways—rhetorical devices that confirm the contestants as celebrities and possibly manufactured authenticity.

Gamson (1994) stated that as audience members grow more active, the more easily they recognize the manufacturing of celebrity. The *American Idol* audience is extremely active, consistently voting, haunting message boards, and reading and reporting assorted show gossip. Even those who aren't actively communicating online are most likely discussing the show at work or home with friends, co-workers, and family. The more knowledge the audience members have, the more aware they are of the manufactured nature of celebrity (Gamson, 1994). This construction is mitigated by the contestants' ostensible ordinariness, revealed through exposure to "behind the scenes," or seeing the contestants' emotions and hard work backstage. Reality television, particularly *American Idol*, clearly demonstrates this cause and effect.

Hill (2005) argued that most audience members expect performance/acting, yet continue to judge authenticity despite this probability. In the context of a reality show, realness is often considered a corollary of honesty; honesty is considered a prerequisite of being on camera and as a characteristic of being authentic. Authenticity is therefore perceived through the emotions of the participants, which are hard to falsify. Audiences search for the moments of real emotions, "moments of authenticity," glimpsed through the performative façade (Hill, 2007). These moments of authenticity are when apparently genuine emotions such as laughter, tears, or anxiety are revealed.

In Hill's (2007) discussion of *Big Brother* as a gamedoc (a combination of a documentary and a game show), she noted possible reasons that audiences watch reality TV include voyeuristic thrill, involvement through interactivity, enjoyment of factual entertainment, and as social currency or something to talk about. Hill found that "viewers prefer informative, behind-the-scenes factual entertainment" (p. 472). Hill's quantitative and qualitative study, conducted in the summer of 2000, found that all ages, races, and genders watch reality TV (or as she calls it, "factual entertainment," also encompassing true-life crime programs). She discovered that audiences expect participants on reality-TV programs to overact and perform for the cameras. In spite of this, they also expect that at some point the participants will break down and act according to their "true" selves (Hill, 2007). These are the critical moments that viewers desire. Viewers are searching for authentic-appearing emotional moments from contestants, regardless of the contrived situations. Audiences engage with reality TV by "negotiating particular frames surrounding 'performance' and 'authenticity' in an otherwise 'constructed' televisual environment" (Pullen, 2004, p. 215). Through this negotiation, audiences achieve a complex understanding of the characters, plot, and conflicts within the programs.

Another way of creating authenticity is humor. Humor provides authenticity amidst acknowledgment of manufacturing. An ironic or self-deprecating joke seemingly provides insight into a celebrity's true nature (Gamson, 1994). Moreover, irony allows a connection between the celebrity and the audience—they, too, get the joke. A celebrity making fun of his or her fame helps maintain the "notion that fame is rooted in character traits, that admiration of celebrities is grounded in merit" (Gamson, 1994, p. 54). For example, Simon Cowell's banter with host Ryan Seacrest often made fun of Simon's constructed mean-judge image, allowing Simon to laugh at himself. He acknowledged his mean image, winking to the audience as he perpetuated it.

DEMOCRATIZING FAME

Celebrities are extraordinary people who represent what an average person can aspire to (Cathcart, 1994). Stars "epitomize the potential of everyone in American society" and, through doing so, represent the "promise of the entire social system to be open to these moments of luck" (Marshall, 1997, p. 9–10). Just as in a democracy anyone can be elected by their peers to hold political office, in the constructed-fame narrative anyone can become famous through media maneuvering and validation by peers (Rojek, 2001).

Reality TV epitomizes the democratic concept that fame is possible for everyone. Couldry (2003) noted that ordinary people are now prevalent throughout television. Dovey (2000, p. 83) described this as "reality TV as empowerment" or the idea that new, seemingly average people are added to the public sphere. Moreover, the ordinary experience depicted on television is more akin to the actual off-screen reality (Turner, 2004). Average people are represented on television now in "multiple versions of class, gender, sexuality, and ethnicity" (Turner, 2004, p. 83). The positive byproducts of more ordinary people being on television are "the openness, the accessibility, the diversity and the recognition of marginalized citizens' rights to media representation" (Turner, 2004, p. 83). Bazalgette (2001) also lauded reality TV for its democratizing effects, specifically citing the representatives from different socio-economic classes often present on *Big Brother* (UK).

Hope for fame is analogous to a desire for an upward move in social stature. In the realm of reality TV, this rings exceptionally true. The average-people celebrities of reality TV often deem themselves lucky to be on the show (Andrejevic, 2004). The reality-TV participant commonly hopes that, through the exposure of his or her life on TV, renown can be parlayed into actual fame. On *American Idol*, contestants aim to win a recording contract and consequently success in the music industry.

Holmes (2004b) asserted that *Pop Idol* embraces a narrative distinct from other reality programs: the traditional quest for success through talent. The program, however, also engages stardom discourses, guaranteeing fame through talent. The winner and contestants of *Pop Idol* are emblematic of the mixture of discourses of the music industry (bland music, appealing to younger audiences), pop music (including the star as an authentic persona yet also carefully styled), and audience interactivity (Holmes, 2004b). Holmes (2004a) claimed that in most reality TV, particularly *Big Brother*, the narrative of hard work or talent leading to success is missing. Accordingly, Gamson's (1994) narrative of constructed fame, lacking work or talent, resonates for most reality TV (although not, notably, *American Idol*). Holmes (2004a) argued that on typical reality-TV programs, the merited rise to fame has been replaced by an increased amount of authenticity, or realness, in the participants. Authenticity supplants talent. However, the success narrative may still be present in the "gamedoc" genre of reality TV (Holmes, 2004a). Indeed, the many versions of *Pop Idol*, *American Idol*, and *The X-Factor* all rely on the narrative of merited success as a basis of the programs. At the same time, authenticity is essential for reality-TV contestants to be viewed as deserving of success.

Evans (2005b, p. 15) claimed that within *Pop Idol* "celebrity is a positive force . . . it represents the power of the individual based on characteristics that are unique to that person alone; it therefore represents equality." People become famous for who they are—their personalities, flaws, and values, as well as singing ability. Contestants on *Pop Idol* "make a virtue of their weaknesses and reinvent their identities, finding a distinctive image that defines them" (15). Programs such as *Popstars*, *Pop Idol*, and *American Idol* provide a link between authentic talent and manufactured celebrity (Holmes, 2004b; Turner, 2004). Through the polished image and a modicum of authentic talent, a normal person becomes a celebrity, symbolizing the democratization of fame. The program represents the contestant as authentically normal and authentically talented, usually both intermingled. The contestants are discovered through the program and simultaneously groomed as celebrities.

CELEBRITY AS SOCIAL CONTROL

The dual narratives of fame, merited and constructed, can be construed as frequently overlapping and affecting each other. While seemingly oppositional, in practice they often enhance one another. Media frequently promote the idea of a natural-born star merely waiting to be discovered through a lucky break. Following the Frankfurt School, this conception of fame can be considered a form of social control. Belief in the concept of democratic celebrity is parallel to a belief in capitalism. Celebrities symbolize the ideals of "heroic individualism, upward mobility, and choice" and thus are a means through which "capitalism achieves its ends of subduing and exploiting the masses" (Rojek, 2001, p. 33). The narratives of fame convince people that anyone can be famous and simultaneously that celebrities are special. Nevertheless, celebrities are constructions: "fabrications designed to enhance the rule of capital" (Rojek, 2001, p. 33). Thus, the more money people spend and the more they continue to believe in a celebrity's fame, the more they support the popular political and economic ideology. Belief in the star system and the equal possibility of fame buttresses capitalism through ideology and consumerism. The culture industry is a key part of capitalism.

Marshall (1997), in *Celebrity and Power,* emphasized that celebrity performs a political function. He explains that celebrities represent capitalistic ideals of consumption and definitions of beauty. Fame may be democratic, but truly functions to hegemonically maintain the status quo of the already established power structure. The adulation of a celebrity guarantees that the extant system, and those who hold power within it, will remain. As such, the "system of veneration . . . is more important than what any one of the individual celebrities represents" (Marshall, 1997, p. 11).

Celebrities and the systematic creation of celebrity help to maintain the current social structure in other ways as well. Audiences may be interested in celebrities in order to help understand the current social system (Dyer, 1986/2004). Celebrity provides distinction for certain people and privileges certain voices over most others. Simultaneously, in the idea of celebrity as social control, celebrity celebrates false value—celebrity is purely based on image, even if audiences may disagree (Marshall, 1997). While celebrities are manufactured, audiences must consume them for celebrity to exist. In effect, the celebrity holds the power of the audience in the public sphere, which is readily apparent in political figures and the power they hold during elections (Marshall, 1997). This power, though, is not employed in the empowerment of the masses, but rather is used for the maintenance of the current social system. As Marshall (1997, p. 214) stated:

> Entertainment celebrities, like political leaders, work to establish a form of cultural hegemony. The meanings of masculinity and femininity, the meaning of family, and the definition of common cultural identity are the various territorial domains upon which popular cultural celebrities navigate in their formation of public personas.

Celebrities communicate "directly or indirectly, particular social values, such as the meaning of work or achievement, and definitions of sexual and gendered identity" (Evans, 2005a, p. 2). For example, narratives of sports celebrities emphasize attributes such as order, compliance, self-discipline, and training (Rojek, 2001). In contemporary times, Marshall (1997, p. 93–94) averred, "celebrity has provided a way in which the discourses of individualism, freedom, and identity have been articulated in modern society." American stars have come to embody the varied discourses of these American concepts. Within *American Idol*, the celebrities also embody varied ideas and meanings.

In addition to creating celebrity, through the planning, editing, and acting of reality-TV shows, certain stereotypes are perpetuated and ideologies endorsed. Stephens (2004) maintained that some reality shows adhere to patriarchal, familial, and standard gender norms. *A Wedding Story* and *A Baby Story* feature familiar rituals and recognizable categories of people, including traditional concepts of family and motherhood, replete with quintessential happy endings. Pullen (2004) observed that gay men are repeatedly positioned as outsiders in many reality-TV programs and only accepted when embracing heterosexual norms, such as traditional conceptions of love, fidelity, and marriage. Indeed, many reality-TV shows correspond to societal standards, encouraging adherence to unchanging society standards. Not challenging the status quo is mute endorsement of it.

While mass media retain primary importance in the creation of celebrity, the audience has some agency in celebrity image negotiation (Marshall, 1997). However, while some celebrities may criticize discourses of power, acceptance by an audience ultimately leads to these celebrities' co-optation by the establishment, negating their subversive influence (Marshall, 1997). A celebrity such as documentary filmmaker Michael Moore may criticize the existing status quo. However, making deals with large conglomerates for film distribution or appearing on network television to promote films may diminish the effect of his message. Working with the establishment mitigates his criticism of it.

Celebrities also "contribute to the cult of distraction that valorizes the superficial, the gaudy, and the domination of commodity culture" (Rojek, 2001, p. 90). This works to diminish the significance of high culture, community involvement, and political awareness. Celebrity and distraction are "a means of concealing the meaninglessness of modern life and of reinforcing the power of commodity culture" (90). Therefore, according to Rojek (2001) celebrity helps obscure important aspects of off-screen life, such as political issues, while simultaneously encouraging materialism, and in turn facilitating capitalism.

In an early work on celebrity, Alberoni (1962/1972) contended that stars, while an elite group, are ultimately a "powerless elite." They have little institutional power or influence. In Alberoni's view, politicians and celebrities inhabit two completely different spheres—celebrities should not have any sway over politics. This viewpoint, while valid in that politicians and entertainment celebrities may (or should, at least) occupy completely different roles, is often invalidated today by the increasingly amplified political vocalness of celebrities. Celebrity endorsement plays a role in politics. During the 2008 presidential campaign, Oprah Winfrey's endorsement of candidate Barack Obama was viewed as a significant accomplishment for his campaign (Zeleny, 2007). Additionally, a positive effect of celebrity is endorsement of charitable causes (Rojek, 2001). An industry has emerged that matches celebrities with potential charities they might be interested in representing (Traub, 2008). While celebrities may actually be altruistic, the positive publicity earned also benefits their images.

AUTHENTICITY, CELEBRITY, AND REALITY TV

Turner (2004) suggested that discourses of possibility—the idea that celebrity can be easily gained—fuel reality-based entertainment. The thousands of people that audition for *American Idol* each year support this claim. Accordingly, the rise in popularity of reality TV has created a plethora of new reality-TV-created celebrities (Holmes, 2004a). Reasons why and how reality TV may create celebrities include the cheapness of production costs, "the hybrid nature of reality TV (the blurring of factual and fictional forms . . .)" (Holmes, 2004a, p. 114), the focus on characters (Dovey, 2000), and the fascination with celebrity (Gamson, 1994).

The people who participate in reality-TV programs are typically thought of as desirous of fame and thus in conflict with the notion of being "real."

They are not "real" people; the participants are wannabe celebrities (Holmes, 2004a). These people perform for the cameras and cannot be authentic. Other authors contend that reality-TV participants are average people (Dovey, 2000; Turner, 2004). Thus, there are supposedly two categories of people on reality TV: authentically average people, representative of the typical population or people who act authentic in order to become famous. These two groups may overlap—authentic non-acting people can also want to be famous. A desire for fame need not preclude the ability to show "moments of authenticity" as Hill (2005) labeled seemingly real emotions. It is difficult to generalize all contestants/celebrities. However, willingness to participate indicates complicity with the publicity that transforms participants into celebrities. Accordingly, Bazalgette (2001) asserted that the desire to show off is the only difference between reality-TV participants and people not on TV.

Whereas talent previously helped produce a celebrity, for instance, through acting or singing skill, it no longer is deemed essential, except in programs like *American Idol*. Authentic normality is required to become a celebrity. An example of this phenomenon may be found in the example of Trista Sutter. Initially a loser on *The Bachelor*, her likeability and success with audiences garnered her a new reality show, *The Bachelorette*. After America watched her find love, she was paid a reported one million dollars to turn her wedding to Ryan Sutter, the winner of *The Bachelorette*, into a reality program that drew over 16 million viewers (Wang, 2004). She also sustained her celebrity with appearances on other reality shows such as *Fear Factor* and *Dancing with the Stars*. Sutter frequently appears in entertainment magazines and shows, appeared in the "Got Milk?" dairy campaign and endorsed KFC. Currently, she is an online columnist for the popular website iVillage and uses her celebrity to sell handbags of her own design. The birth of her and Ryan's son, Max, spurred another round of magazine interviews. A few months later, stories about her weight loss appeared along with pictures of her in a bikini. In the fall of 2008, stories began to surface of a second pregnancy and by summer 2009, new baby girl Blakesley was featured in *US Weekly* and *People* magazines. Sutter, working with the viewing audience, transformed herself from an ordinary person into an extra-ordinary celebrity. Her normality is celebrated and has turned her into the new type of celebrity—extra-ordinary, instead of extraordinary. Her lack of inherent talent or skill is replaced by publicity, venerating her ordinariness. Trista illustrates Boorstin's (1961/1992) tautology: familiar for being familiar, famous for being famous. The authentic normality that Trista represents is a requirement for would-be *American Idol* celebrities as well.

On *American Idol,* authenticity is composed of three components. The first considers emotions as symptoms of the authentic or true self. The second refers to actual talent, primarily musical ability. The third is loyalty to archetype, or creation of self-image, determined by every action and performance of a contestant. Emotion, talent, and archetype are part of the naked self of the performer—acting cannot create true emotion or talent. *American Idol* contains both the traditional merited narrative of success combined with reality TV's need for authenticity. *American Idol* contributes its focus on merited success and authenticity to American culture and mythology.

MYTH

Media aid in the construction and articulation of American mythology. Whereas ideology may encompass ideas pertaining to government, economics, and institutional concepts, mythology pertains to cultural values and norms. The two are similar and intertwined—myths are often part of and support ideologies (Fiske, 1987). In entertainment television narratives, symbols and codes contribute to the overall creation of mythology, informing and naturalizing culture for audiences.

Myth is a system of communication that is a message itself (Barthes, 1957/1995). Therefore, myth is not defined by "the object of its message, but by the way in which it utters this message" (p. 109). Along those lines, "myth is a form of speech . . . marked by definable narratives, familiar, acceptable, reassuring to their host culture" (Silverstone, 1988, p. 23). Myths are a system of meanings, usually straightforwardly accepted by audiences, transmitting values and concepts (Strinati, 1995). Wright (1975, p. 11) considered myth to be "an abstract structure" containing "symbolic content through which the social structure is applied to contingent, socially defined experience."

Through structure and content, "myth uses the language of other systems, be they written or pictorial, to construct meanings" (Strinati, 1995, p. 113). Within myths, meaning of elements may be obtained through intertextual references. Myth is a socially powerful narrative (Sillars and Gronbeck, 2001) and a means through which culture is transmitted (Silverstone, 1988). Furthermore, myth "transform[s] history into nature" through both telling and understanding (Barthes, 1957/1995, p. 129). Therefore, myths naturalize stories, making events customary, understandable and expected. They give "historical intention a natural justification" (Barthes, 1957/1995, p. 142). Myth defines and maintains commonsense reality (Silverstone, 1988) and has the power "to empower positive and negative values and behaviors" (Sillars and Gronbeck, 2001, p. 179). Thus, myth informs people about humanity, both through form and content.

Lule (2001, p. 15) defined a myth as a "sacred, societal story that draws from archetypal figures and forms to offer exemplary models for human life." In news today "myths are archetypal stories that play crucial social roles" and as such, are "an important way a society expresses its prevailing ideas, ideologies, values and beliefs" (p. 15). Thus, myths help construct society through fostering acceptance of social structure, values, and dominant ideology. News and myths help sustain social order (p. 15). Moreover, archetypal stories "offer exemplary models" demonstrating idealized ways to act. Within myth research, investigators look for reiteration, patterns, and archetypes that may provide guidelines for audiences.

American Idol contributes to the maintenance and adaptation of American cultural myths. It is a medium of mythology and ideology transmittal, providing stories that naturalize aspects of culture, including hopes and dreams. Narratives, archetypes, and symbols provide meanings for *American Idol*. Moreover, intertextualities affect meaning as well (Fiske, 1987). How the program interacts with other television shows and other media affects meaning. This study addresses this by examining popular press articles about the program. Thus, studying *American Idol* permits examination of embedded American discourses and ideals, ideology and mythology, factors of what makes American culture uniquely American.

METHOD

For this book, I used a three-pronged approach to study celebrity, collective identity, and American themes in the program *American Idol*. First, I conducted a discourse analysis of the first seven seasons of *American Idol*, focusing on the contestant narratives of the top twelve contestants from each season. Throughout the show, a narrative is constructed for each contestant, explaining why he or she deserves to be the American Idol. Every image and word shown or uttered on the show was considered evidence. The first season consisted of twenty-five shows, the second had forty-two episodes, the third and fourth seasons were forty-three episodes each, the fifth and sixth seasons were forty-one episodes each, and the seventh season had forty-two episodes. In total, the research spanned 277 episodes. The results episodes were one hour in length, while the performance episodes were typically two hours long. Throughout each episode, I noted repetitive occurrences, quotations, and visual components. (Throughout this book, I refer to contestants by their first names, as the show typically does. This also contributes to the portrayal of the contestants as average people, as discussed later.)

Within the contestant narratives, I specifically examined details of a rags-to-riches journey, potential archetypes, and discourse themes concerning race, class, or gender. I then examined recurring character archetypes communicated by the show through visual cues of physical appearance, musical genre preference, and background information. The metamorphosis of each contestant over the course of the program from being simply a talented average person into a celebrity is examined through the analysis as well, particularly notions of authenticity versus construction. Within the discourse of the show, statements attesting to contestants' star quality and transformation of physical appearance were noted as part of the confirmation of celebrity-status process. I also paid particular attention to common contestant background elements. Everything about a contestant adds to his or her identity—therefore, the examination of discourse surrounding him or her illustrates how the program fashions concepts of America, American identity, and celebrity.

The second portion of the study expands the discourse analysis to include *American Idol* media coverage. This is a press coverage discourse analysis of the 2002 first season and the 2008 seventh season. The qualitative study of press coverage includes most major newspaper and magazine coverage, as well as transcripts of television and radio news programs, and episodes of programs such as *The Tonight Show* and *Good Morning America.* Excluded articles and transcripts are those that focus on ratings, recaps, or are only tangentially related to the show.

The third portion of the study focuses on the interactive characteristic of *American Idol.* During the 2008 season of *American Idol*, I spent at least four hours per week in the discussion areas on the official *American Idol* message boards, and collected transcripts of online discussion about the show. Each season has a dedicated message board, with eight categories of discussion: general *American Idol* discussion, rules, FAQs (frequently asked questions) and announcements, your recaps, judges, Ryan Seacrest, favorite idol photos and video, games, and off-topic discussion. Within each category there may be thousands of discussion threads. I focused primarily on the active threads, and particularly recorded any indicators of community formation, maintenance, or termination. The analysis revealed that fan communities and subsequent collective identities form around the show itself and certain contestants. I searched for discourse cues: repetition, exaggeration, and positioning of fans in the community through word selection.

Moreover, I noted what participants found most important about the show, their likes and dislikes, pertinent thoughts they chose to express, and overall themes and consistencies. Following Andrejevic's (2004, p. 123) study of *Big Brother* message boards, by "juxtaposing a critical analysis of the shows with the responses of fans. . . . some fruitful patterns for thinking about the social significance of the genre . . . emerge[d]." In the analysis of this study, I was able to extrapolate ideas about collective identity through online discussions of aspects of the show, including dislike or like for particular contestants. Aversion to certain contestants and groups exhibited the us-vs.-them social-identity theory (Tajfel, 1986) phenomenon of reinforcing community identity through common distaste.

DISCOURSE ANALYSIS

Fairclough (2003, p. 3) stated that discourse is a "particular view of language . . . as an element of social life which is closely interconnected with other elements." Similarly, simpler definitions are "particular ways of talking or writing in specific contexts" or "almost any subcategory of English language use" (Sillars and Gronbeck, 2001, p. 148). Varied "discourses are different ways of representing aspects of the world" (Fairclough, 2003, p. 215). Ang defined discourses as "situated practices of functional language use and meaning production" (quoted in Strinati, 1995, p. 250).

The term *discourse* may apply to any written or spoken language, and can be extended to visual images and non-verbal communication (Fairclough, 2000). Fairclough alleged that discourse is socially shaped and that discourse also shapes society. This dialectic relationship is "socially constitutive" (p. 309). Through discourse, humans build relationships with each other (Sillars and Gronbeck, 2001). Critical discourse research analyzes language use in reproducing and maintaining social identities and systems of knowledge (Fairclough, 2000). It analyzes communicative events in three parts: texts, discourse practice, and sociocultural practice. A text is a written or oral part of social interaction, the discourse practice is text production and consumption, and sociocultural practice is the context of the event (Fairclough, 2000). Within discourse, "media 'messages' are specific types of text and talk" (van Dijk, 1991, p. 108). Discourse analysis creates a "systematic and explicit account of the structures of media messages" and relates the structural account to "various properties of the cognitive and sociocultural context" (p. 108). Thus, discourse analysis contextually explores concepts within media. Taking all of these scholarly definitions into account, discourse may be characterized as a portion of speech, either verbal or written, generally advocating a particular point of view, and often used to position authority.

There are several discourses present on the show, including one that positions the audience as expert judges. Another discourse constructs an audience receptive toward dominant meanings, for example, the construction of an audience into individual consumers. Commercial discourses attempt to redefine audiences into consumers, receptive of intended commercial messages (Strinati, 1995). Discourse is regularly the site of institutional power operation.

NARRATIVE ANALYSIS

Narrative analysis is a type of discourse analysis. Narratives are considered an integral part of human life, as well as stories used to explain humanity and the world (White, 1981). A narrative is "based on experience, is a product of memory, has a sense of chronology about it, is coherent, and has a central subject" (Sillars and Gronbeck, 2001, p. 218). Narratives may be "the metagenre of expressive culture," originated from tribal rituals and the progenitors of all communication (Turner, 1981, p. 163). They provide context and give meaning to speech (Derrida, 1981). Narratives give meaning by "draw[ing] on the past to make sense of the present and future," providing contextual understanding (Kitch, 2005, p. 2). Thus, narratives demonstrate how to understand society and are vital to human communication (Sillars and Gronbeck, 2001).

As film scholar Chatman (1981, p. 117) noted, "Narrative itself is a deep structure quite independent of its medium." Narrative exists separately from film, TV, novels, etc., yet can be found in all media. No matter the medium of expression, be it film, television, or a human orator, there are two aspects to a narrative: presentation by an actor and reception by an audience (Kermode, 1981). Through comprehension of narratives, reality is constructed (Zelizer, 1992). Narratives are culturally specific, located in accordance with a certain sociopolitical order (Bird and Dardenne, 1988; White, 1981). Accordingly, stories and narratives "are symbolic actions that create social reality" (Sillars and Gronbeck, 2001, p. 215). They do not merely reflect social structure but are a means to learn about culture from those within it (Sillars and Gronbeck, 2001). Narratives also encode values and act as behavioral guides (Bird and Dardenne, 1988). Correspondingly, they connote ideology and present a moral version of reality (White, 1987). Consequently, narratives may be defined as repetitive stories with a definitive structure of sequential elements that aid in the construction and comprehension of human society. Overall, when employing narrative analysis, researchers may look for repetitive narratives and stories; sequential occurrences; and recurring events, morals, or characters.

Narrative devices in reality TV occur in verbal and visual forms (Cavender, 2004). Often, hosts will directly address audiences, including them in the program. Contestants are addressed by their first names, invoking familiarity. In *Big Brother*, a voice-over commentary provides a meta-narrative, explaining events and hinting at possibilities, as well as providing a clear structure for the program (Tincknell and Raghuram, 2004). By adding to the show details, the voice-over concluding remarks at the end of each episode instill narrative closure and provide intimacy between the audience and the participants (Tincknell and Raghuram, 2004). Visually, scenes depict community: on *Survivor*, contestants are placed symbolically next to each other in a tribal setting (Cavender, 2004).

The entire program of *American Idol* itself exists within a narrative: the search for the next American idol. This is established through rhetorical choices in dialogue by the program host and judges and reiterated in contestants' interviews. The process of voting by the audience confirms this narrative, as well as involves the audience. This narrative influences understanding of the show, allowing contestants to try to prove their talent, likeability, and self-worth in a quest to prove they deserve to win.

CONCLUSION

This book focuses on the contestant narratives for each season within the program *American Idol*. Recurring elements and themes in the narratives are explored. Moreover, each contestant's narrative provides contextual meaning for the audience. These narratives over time become familiar to audience members, allowing for preconceived knowledge to influence perceptions. Therefore, the contestants attempt to adapt to the expectations of their chosen narrative. For example, a season-five contestant may try to correspond with a narrative and archetype established by the previous seasons. The narratives and archetypes thus provide a way of understanding the contestants and the show itself, as well as reflect and help create a social construction of reality. Narratives and archetypes are good tools of analysis not only because they provide familiar connotations for the audience, but also because they are identifiable through repetition.

During the discourse analysis of the contestant narratives, I studied contestant archetypes. Archetypes (which persist through time and different cultures) and stereotypes (specific to a particular time and place) are excellent tools for analysis because they are readily visible and clearly provide roles audiences understand.

Thus, this project includes narrative analyses of the top-twelve-contestant stories in the first seven seasons of *American Idol*, a discourse analysis of the 2002 and the 2008 press coverage, and the examination of fan collective identities pertaining to *American Idol* in the official message boards. These three analyses combine into an overall discourse analysis of *American Idol*. At end of the book, I draw conclusions about celebrity, American ideals, and collective identity in the first portion of the new millennium.

Chapter 3

Rags to Riches and the Discourse of Dreams

The most dominant narrative within *American Idol* is the rags-to-riches journey. This narrative begins with a poor, unknown contestant who becomes rich and famous through the program. This voyage is present in many media texts and has been a prevailing American narrative for over a century (Caldwell, 2006). From the immigrant who hopes for a better life to the buxom blonde hanging out at Hollywood and Vine, people search for their own rags-to-riches narrative in America, the land of expectations. *American Idol* continues the tradition of this narrative. In order to establish that the contestants need and deserve the rags-to-riches journey, the show discourse stressed the blue-collar beginnings of most contestants. Many held service-industry jobs, were aspiring musicians, or were unemployed prior to their season. In the poverty-to-wealth narrative, anyone can be the star no matter race or background, as the racial diversity on the show underscores. Not only are the contestants racially diverse—black, white, Asian, and Latino—but also in a true representation of America, the judges are racially diverse as well. The black man, white men, and white women (and later, a homosexual white woman and a Latina woman) also represent the fruition of this narrative and the success realized at the end of the narrative journey. Likewise, in an example representative of many, host Ryan Seacrest once stated that the winner of *American Idol* will "live a celebrity life with fame and fortune" (July 23, 2002). This is the expected fulfillment of the rags-to-riches narrative. The *American Idol* embodiment of the Horatio Alger story illustrates the capitalistic dream, reinforcing the contemporary American ideological status quo. This rags-to-riches narrative is apparent throughout all the seasons of the show through discursive emphasis on promises of success, contestants' lives and humble beginnings, and dream fulfillment.

SHOW DISCOURSE

The original title of the program was *American Idol: Search For a Superstar*. This was the title for the first season; the title was then shortened to simply *American Idol* for the subsequent seasons. Even by the sixth season finale, judge Randy Jackson was still referring to the search for the superstar. He once said,

> You know what is great about this show? What I love about this show is that it is *American Idol*, the search for the next superstar. It is a singing competition bar none. It is the best singing competition ever in the history of television. (May 22, 2007)

While the title may be shortened, the meaning of the longer title still resonates for people within the show, as Randy demonstrated by lauding it five years later.

The discourse of the show repeatedly stressed that *American Idol* is a competition seeking the best of America. For the show champion, this narrative culminates with winning the competition and enjoying a subsequently successful career. As Ryan once said, "It's someone's career, not just a television show" (April 29, 2003). Later, in season three, Ryan pointed out the stakes of the competition as "a future you could basically sum up in three words: fame and fortune" (March 10, 2004). Judge Paula Abdul also repeatedly emphasized the point of the show, saying at one time "It's been a great season and it's time for them to get on with their careers" (May 20, 2003). On the seventh season, as images of Kelly Clarkson appeared on screen, Ryan's voice-over said, "The last time we had auditions in Dallas, we found a twenty-year-old waitress with a dream of her winning. Overnight, she became a global superstar" (January 16, 2008). Within one day of her win, the discourse asserted Kelly was a global superstar.

Therefore, the foreseen end result of the show is a successful music career inclusive of fame and fortune for the contestants, the completion of the rags-to-riches narrative. Ryan's statement on the season-four finale was illustrative of this:

> It is the moment the season's been building to. Just a reminder of what's at stake here: a record contract, a whole group of new friends, and here's something else amazing—waiting for the winner is this Marquis Jet card. It will get the lucky winner access to their (sic) own personal private jet. (May 25, 2005)

The winner of the show received a record contract, and a celebrity lifestyle indicated by the new group of friends and a personal jet.

The show is also inextricably linked to Los Angeles, the site of Hollywood. Hollywood is the center of the entertainment industry and the origination point of many celebrities, including the *American Idol* winners. At the start of the first show, cohosts Brian Dunkleman and Ryan Seacrest stood in front of the iconic Hollywood sign and explained the potential success of the show's winner. They asserted the British *Pop Idol* winner "went on to make over a million dollars and two number-one hit records within three months of his victory" (June 11, 2002).

Associating the show with Hollywood, the iconic Hollywood sign, and the Kodak Theater, home of the Academy Awards, lent credibility to the program and its promise of success for the contestants. All of these physical sites imply through association that the program can actually provide success for the eventual winner. An Academy Award is the highest honor an actor can win. *American Idol* may be the highest honor an amateur singer can win. This connection may have foreshadowed the eventual winning of an Academy Award by an idol contestant, Jennifer Hudson, who is to date the only reality-TV show contestant to win one. During the audition process, those contestants who pass the initial audition receive a golden ticket and hear the phrase "You're going to Hollywood!" The golden ticket may reference the movie *Charlie and the Chocolate Factory*, in which the golden ticket represents escape from poverty and the monotony of daily life. Moreover, the show discourse continually restates the association between Hollywood and the program throughout every season. For example, in season two Ryan said, "We're coming to you live and direct from Hollywood" (April 8, 2003). *American Idol* is the new Hollywood for singers.

A final piece of evidence of the meteoric rags-to-riches rise was proffered several times in various seasons by Clive Davis, the "chairman and CEO of RCA records." On the season two finale, he outlined the potential success of both of that season's finalists, Clay Aiken and Ruben Studdard. Clive Davis said:

> As Simon Cowell has repeatedly emphasized, this competition is all about finding the most talented American idol and the prize is a recording contract from a major label. So we at Jive records, at RCA records, and BMG, we've made every effort to keep the bar up there. For example, we just finished recording Justin Guarani's album. And I'm here to tell you that he's totally risen to the occasion. We've already begun recording Clay and Ruben. And no matter who wins tonight I can tell that in Ruben and in Clay you've chosen the best. Each is the real deal and they're not going to let you down. These artists, once they record, they're thrust into direct competition with Norah Jones, Justin Timberlake, and ultimately our very own Alicia Keys. And so this brings me to the reason I'm here tonight. Last year's champion, Kelly Clarkson, she had her album

> released five weeks ago. And did she ever rise to the challenge. Her album got raves from *Entertainment Weekly*, from *Billboard*, from even *Rolling Stone*. It entered the top 200 Soundscan chart at number one. Her single, which you just heard, 'Miss Independent,' it's exploding, on its own very merits, to the top of every chart in this country. So tonight, all you fans across America, your champion idol, Kelly Clarkson, she's more than meeting the top competition. Her album, in just five weeks, has now sold over one million, three thousand copies. So, please celebrate with me, as I present to Kelly tonight the very special RIAA certified platinum award for her album, *Thankful*. (May 20, 2003)

Clive Davis detailing the accomplishments of winners, particularly the flourishing Kelly Clarkson, validated the show. Davis also repeatedly referred to "you," the audience, as the experts who chose the winner, Kelly Clarkson, thus positioning the audience as experts. Through this speech, audiences were able to see *American Idol* really could create superstars and provide the magical ending to the rags-to-riches narrative. Kelly Clarkson's accomplishments, according to Davis, proved that the show delivers on its promises.

Blue-collar background

While the show provides the ending of the rags-to-riches narrative, the contestants must furnish the lowly beginning. Through the show discourse, audiences learn that most contestants grow up in blue-collar homes, most hold service-industry jobs, and many are single parents. On the first episode of the program the two cohosts, Ryan Seacrest and Brian Dunkleman, explained the nature of the rags-to-riches narrative.

> Ryan: "This is the Kodak Theater in Hollywood, home of the Academy Awards. And possibly the most televised theatre in the world."
> Brian: "Three months from now, live on this very stage, an as yet unknown talent will be launched into superstardom."
> Ryan: "We don't know who that is yet. Right now, they could be parking cars or even waiting on tables, who knows."
> Brian: "What we do know is, at the end of the summer, that person's life will change forever."
> Ryan: "Because you at home decide who will become the next American idol." (June 11, 2002)

According to Ryan, the unknown talent presumably is not an attorney or an executive. Those people have already achieved success. This program is for blue-collar, working-class people, searching for a chance to escape a dead-end job and wind up being a peer of Mariah Carey. A valet or waitress, with hard work, talent, and the help of the program, can become famous. Whoever the winner is, "that person's life will change forever" as presumably he or she is "launched into superstardom."

Just as in off-screen American culture, on *American Idol* contestants are defined by their jobs prior to appearing on the program. By season eight, as people auditioned, a banner appeared underneath their faces, announcing each person's name, age, and job. Occupations before appearing on the show are repeatedly stressed in interviews, packages, and the overall show discourse as a way to identify the contestants and explain who they are. In an example, Ryan explained the season-seven final three contestants: "Last year, three people took a chance and auditioned: a high school student, an actress, and a bartender. One of them is on the brink of superstardom. All of them are at your mercy. This is *American Idol*" (May 13, 2008). Of the three finalists, not one had a viable middle-class career. On *American Idol*, contestants' careers (or lack thereof) indicate class. If a contestant grew up in a middle- or upper-class home, that portion of his or her life is downplayed or omitted. First season runner-up Justin Guiarini, for instance, was more frequently linked to his past as a door-to-door salesman than his childhood in a middle-class suburb. All contestants are portrayed as working class, and in need of the show-provided rags-to-riches journey.

For example, season-two winner Ruben Studdard was unemployed, lacked a college degree, and lived in Birmingham, Alabama. The socioeconomic class he grew up within was never explained. Ruben wanted to be a professional singer. The *American Idol* contestants desire lucrative careers as entertainers, most likely accompanied by the trappings of celebrity and wealth. *American Idol* capitalizes on this desire, using it as the overall theme of the program. As Ryan Seacrest said at the season-one finale: "Now, Kelly was a cocktail waitress, Justin was a door-to-door salesman. One of them is about to be a superstar" (September 4, 2002). Both Kelly and Justin were positioned as working class.

The contestants themselves often discuss their working-class backgrounds, establishing their class positions. Season-four contestant Scott Savol said:

> I had a few odd jobs after high school, back in the 90s, barbeque spots, pizza shops, restaurants. No, but the whole time I was doing that I was dreaming of being in the music industry. You know, singing, which is what I love to do, so I decided to put those things aside and really pursue my singing career. And here I am now. (March 29, 2004)

While Scott may have held those jobs, it sounded as if he was unemployed at the time of his audition, although hoping for a singing career. Chris Daughtry also held a job while dreaming of fame. He said at his initial audition: "This would change my life tremendously. I wouldn't have to work a day job anymore. I'm a service advisor for a dealership in North Carolina" (January 18, 2006). Similarly, Bo Bice worked at a guitar shop, Vonzell Solomon delivered mail for the U.S. postal service, and Kellie Pickler was a roller-skating waitress. In season seven alone, contestant Ramiele Malubay worked as a restaurant hostess, David Hernandez was a waiter, Brooke White was a nanny, Michael Johns taught tennis, Chikizie was an airport security screener, and Carly Smithson and David Cook were both bartenders.

Elliot Yamin was a season-five contestant who also held low-level jobs. He discussed with Ryan his background:

> Ryan: "Before *Idol* took over your life, what were some of your jobs, Elliot?"
> Elliot: "Oh my goodness, we'd be here all night if we had to talk about that."
> Ryan: "So, you couldn't keep one?"
> Elliot: "You know, I worked fast food, corporate settings, I've done just about everything." (April 11, 2006)

The corporate settings jobs are never mentioned again. They may have been glimpses into a more complicated story. However, beyond simply being portrayed as blue-collar, many of the contestants had other obstacles to overcome. Pity-inducing stories were frequently detailed on the show.

Hard-luck Stories

Contestants tell stories that range from heartbreaking to laughable. Typically, most background stories try to evoke compassion for the contestants. And indeed many of the hard-luck stories which were emphasized over the first seven seasons were pathetic, such as Kellie Pickler growing up while her father was in prison, while some were less pitiful, like Anwar Robinson's choir's bus breaking down on the way to sing for President Bill Clinton. Many health problems were described, such as season-four contestant Anthony Federov's tracheotomy as a baby, which left even his speaking ability in doubt. Season-seven contestant Amanda Overmeyer survived a car accident that occurred just before Hollywood week. Kristy Lee Cook had to sell her favorite horse in order to raise the money to compete. Kristy Lee Cook was positioned as poor, which ignores the fact that *she owned a horse*. Whether truly horrific and sad or merely circumstantial, the contestant sob stories were an integral part of explaining problems the contestants surmounted on their journeys from rags to riches. Whether or not these stories are true is insignificant. They accomplish the goal of positioning the contestant as working class and deserving of the rags-to-riches journey. The sob stories also serve as a direct tie between ancestral programs such as *Queen for a*

Day and *Strike it Rich*, in which participants related their sob stories on camera in order to win prizes.

Season-five contestant Elliot Yamin explained the health troubles he overcame:

> Believe it or not, I actually have a ninety-percent hearing loss in my right ear. I had a lot of infections growing up when I was a kid. No need to feel bad, my friends back home are always jokin' with me. It's almost like playing a game of telephone; I would probably lose every time. So, I think I'm just going to stick to singing. (March 8, 2006)

Paula was touched by Elliot's sob story of health issues and said, "Elliot, you are absolutely phenomenal and amazing. Hearing, like, the blueprint of your life and what you've had to overcome in obstacles even makes it a richer story for your background. I'm beside myself" (March 8, 2006). In addition to the hearing loss, Elliot also explained he had diabetes and strong allergies. However, Elliot managed to prevail over his health problems, compete in the show, and finish in third place in the fifth season.

Some contestants had to contend with more pervasive and societal problems. Season-three contestant George Huff spoke about growing up in a poor neighborhood in New Orleans. He said:

> My childhood, growing up in New Orleans, we grew up in the projects. We lived in the projects for maybe eight to ten years. You know, you have your hardships and everything, but you pick yourself up and you just find a place of peace in the situation that you're in. One of those places that I found was in my music. (March 23, 2004)

George's story was sad but uplifting, a repeated theme. Stories portraying past distressing times and ending in a better present were common. Season-seven contestant Syesha Mercado's drug-addicted father appeared several times, beginning with her first audition. The man had gone through rehabilitation, but could still barely speak. He did manage to address Syesha's success, saying "I struggled through drugs and alcohol for years and this is a natural high for me. It gives me a purpose and a reason to stay clean" (May 14, 2008). Syesha's unfortunate background of poor parenting due to her father's drug addiction provided the requisite rags portion of her rags-to-riches journey.

However, Kellie Pickler had the ultimate sob story. She explained through tears,

> I live with my grandfather. My mother left me when I was two years old, which is devastating for a child, because you wonder, why didn't your mother want

> anything to do with you? My father's been in and out of prison pretty much my whole life and uh, he's a drug user. It's hard to see somebody that you care about, especially your own father, locked up in prison. I have nothing to go back home for. (January 24, 2006)

For Kellie, appearing on *American Idol* was most definitely a rags-to-riches story, enhanced only by her chart-topping and award-winning success after the show. Moreover, her post-show breast augmentation, particularly evident on her later return visit to the show, confirmed her ending of the rags-to-riches journey. Her purchased breasts represented her new wealth and also typical celebrity beauty standards. The hard-luck sob stories, especially in conjunction with blue-collar jobs, firmly establish that the contestants are in dire need of the fairy tale rags-to-riches voyage.

CONTESTANT DREAMS

The show discourse also repeatedly referred to the contestants' dreams, particularly their common dream of success. Success, to some, may have meant recording contracts, celebrity status, and money. To other contestants, success was simply being on the show. Appearing on the show was a modest goal in comparison to fame and riches, and this humble goal may have helped to appeal to voters. The discursive use of the word *dream* and the connotations associated with it support the concept of the rags-to-riches narrative. The contestants' dream was essentially to attain the end of the rags-to-riches journey.

To begin with, the audition process illustrates that many people will do anything to be on television and briefly be famous. People wait for hours just to perform a joke song or humiliate themselves in an attempt to earn fame or notoriety. The program elects to show contestants who demonstrate vehement emotions upon not being chosen. Clearly, this makes for interesting television. The success of the early programs also indicates that emotions are what audiences prefer to see. Many of the auditioners seem very emotional, sad or angry when they don't make it to the next round. In a time where scores of people on reality TV become famous despite having little apparent talent, many people think they too deserve to be on television. Seeing those who fail to continue to the next round also reinforces that those contestants who do succeed are truly worthy. On the other hand, a considerable number of the contestants stated that simply appearing on the show is an achievement of their dreams. Judge Simon Cowell disagreed that the accomplishment of being on the show is enough of a goal. He once said,

"This is a winner's competition. I disagree about everyone being winners in the top four. Everyone should want to win this competition" (May 11, 2004). Nevertheless, being on the program, being viewed by millions of people is an accomplishment of a dream: fame created through singing. Singing on *American Idol* means the contestant is famous and essentially is a professional singer, at least while on the show.

Dreams

The contestants and the show discourse repeatedly referred to their dreams. This discourse of dreaming suggests a popular dream, or hope, is to be famous. An example of how the show promotes the concept of a dream was visible on the season-six finale. Ryan said "Two very different people, with two very different styles, chasing one unforgettable dream. And you at home will make that dream come true" (May 22, 2007).

Tamyra Gray, on the top-five contestant episode, stated, "To be in this place right now, it's a dream, it's really surreal." For her, after a lifetime of trying to make it in the entertainment industry, simply appearing on the show was an achievement. After so much rejection "When someone finally says yes, you're like, are you for real?" She said, "It's very much a Cinderella story" (August 13, 2002). Tamyra, with the help of the show and her voice, magically turned into a well-known entertainer.

In another example, first season cohost Brian Dunkleman said, "Kelly realized she'd been given an opportunity to fulfill her lifelong dream and nothing was going to stop her now." Kelly responded, "If you would have said two months ago I'd be performing on national television I would have probably laughed in your face" (August 13, 2002). The dream of fame and stardom is made to appear very far out of reach for the contestants, enhancing the star-quality appearance of the finalists. And, as Brian stated, "In less than four months, the singers had gone from virtual unknowns to household names" (August 13, 2002).

Season-three contestant George Huff said on the evening the top twelve contestants performed, "My experience on *American Idol* has been my dream. All of the singing I've been doing over the years—I'm finally at the place where I've dreamt to be. I can't believe it" (March 16, 2004). Brooke White echoed this concept in season seven. Randy asked her, "I was sitting here, imagining you at home, you practicing this as a kid, growing up thinking one day I'm going to make it, I'm going to get on some show. This is kind of a dream come true for you, right?" Brooke answered, "In a big way" (March 11, 2008). The judges and host also highlight and contribute to the dream discourse. Everyone on the show, contestants, hosts, and judges, assume that audiences will understand what their dreams may be: being on television, singing professionally, and attaining fame and wealth.

Immigrant Dream

The rags-to-riches, Horatio Alger narrative has a unique application for one particular group of people—immigrants. This tale of a journey and hard work is found not only in media, but also often in individual Americans' family histories. Over the years, *American Idol* featured several immigrants and first-generation Americans, including Camile Velasco, Leah Labelle, Anthony Federov, Carly Smithson, Michael Johns, and Chikizie Eze. Through their discourses, it was plain that the rags-to-riches American dream resonates for everyone, particularly immigrants, who seek fame and fortune in addition to freedom and safety.

American Idol season-three contestant Leah Labelle was a first-generation American. She related a cold war tale of escape:

> My parents were in a music group together in Bulgaria and they took the opportunity to get visas and go on tour to defect to America. When my parents did escape, it was Communist and it was very difficult to even leave the country. They left everything they knew and they loved. Didn't know any language but Bulgarian. (March 16, 2004)

Leah's mother added:

> It was difficult to leave our families knowing that we wanted to have our own family. We were planning to have children where we can give them more opportunities. I'm very proud to be an American and if Leah becomes an American idol it will be the biggest blessing. (March 16, 2004)

Leah concluded, "For me, this is the American dream" (March 16, 2004). Her participation on the program possibly represented a culmination of her (and presumably her mother's) American dream: success, fame, and fortune achieved through a democratic system, visible both on the program and in the American government.

The next season, fourth-place finisher Anthony Federov expanded the immigrant narrative. He explained:

> I moved to the United States when I was nine years old. I came from the Ukraine. When I came here, I didn't know how to speak English, so me being the Taurus that I am, and me being as persistent as I am, I told myself that I'm going to learn the language and I mastered it. And here I am today, on *American Idol*, singing in English, speaking in English, and I'm happy doing it. (March 7, 2005)

Appearing on the show signified his true acceptance in the American community, achieved through his determination to assimilate. His immigration functioned as his uplifting sob story. Anthony's hard work (learning English, assimilating) is integral to his rags-to-riches narrative.

Later in the season, Anthony continued to discuss his emotions about moving to America. He said,

> When I came to America in [19]94, I was nine years old. And you know, when I first came here it's like coming into a whole new, different world. I mean, everything looked different. You know, the cars looked different, the stores looked different, the street signs were different, you know, the people looked different. I didn't know what any of it meant, but I knew that I was dying to be a part of it. (March 29, 2005)

While discussing his initial time in America, he omitted saying anything about the Ukraine, especially any longing for his birth country or homesickness. He seemed happy to have left the Ukraine and ready to embrace being American.

While discussing his song choice of "Climb Every Mountain" from *The Sound of Music*, Anthony stated, "When I first heard it, I felt touched by it because it deals a lot with determination and achieving a dream, which is exactly what I'm here to do" (April 5, 2005). His mother explained through a thick accent that "He's our dream. His dream, our dream. And I can't believe our dream came true." His father said, "So proud of my son. He's not just a dreamer, he's dream maker" (April 26, 2005).

Upon his elimination on May 10, 2005, Anthony thanked his parents, saying:

> Thank you so much for basically giving me the all-American dream. You guys brought us here for a reason, to make a better life and I'm living it. I'm making all my dreams come true and that would never have happened if it wasn't for you two. So thank you, I love you both and I will make you guys proud.

Anthony implicitly explained that simply having the opportunity to succeed is a crucial part of the immigrant American dream.

Similarly, seventh-season contestant Carly Smithson, originally from Ireland, said upon her elimination, "America is the place of opportunity and the American dream. I feel like this itself is an accomplishment. I feel like I've won already." Paula agreed with her, "You are living the American dream and it's only just begun" (April 23, 2008). The immigrant rags-to-riches dream discourse reinforces that America is the land of opportunity, a place where everyone has an equal chance to succeed. This discourse ignores issues such as xenophobia, racism, sexism, and homophobia. America is idealistically positioned as a utopia, a melting pot, where everyone is welcome. This discourse suggests that anyone in America, including immigrants, can achieve the rags-to-riches dream through hard work and talent.

EVIDENCE: WHERE ARE THEY NOW?

At the end of the *American Idol* narrative, the contestant has achieved his or her dreams and is wealthy, well known, and an accomplished musician. The show discourse asserted that any of the final twelve contestants on *American Idol* could achieve this, even if not the ultimate winner. During season seven, viewers were treated to anecdotal "where are they now" features of successful contestants. The successful contestants demonstrated the fruition of their ultimate rags-to-riches narrative and promoted their careers by appearing on the show. Those contestants who did not proceed to successful careers after appearing on the program were absent from the "where are they now" features. This omission allowed the idealization of the rags-to-riches journey. Moreover, the packages demonstrated that anyone with talent, regardless of race, can achieve success. These evidentiary packages proved that *American Idol* did fulfill the rags-to-riches narrative—at least for some contestants.

In a package featuring season-five contestant Kellie Pickler, she spoke backstage before performing at a concert and explained her transition:

> I was the roller-skating waitress. You have the footage. My life has done a complete one-eighty, all for the best. This is what I signed up for. This is all I've ever wanted since I was a little girl. Very, very great and blessed to be here. I don't really feel any different as a person. I feel the same—I just have a waaay better job. . . . There's no greater reward than waking up and singing to thousands of people. (March 19, 2008)

The following week, a package highlighting season-two second-runner-up Kimberly Locke aired. Kimberly was pleased to report that she had a record deal before her *American Idol* summer tour was over. Also, she had since lost forty pounds, opened a restaurant outside of New York City, and had a second solo album to promote. All of this occurred because "*American Idol* put me in a wonderful position to do whatever I wanted to do" (March 26, 2008).

The next week, a trio of contestants all living their dreams in Nashville were profiled: Bucky Covington, Phil Stacy, and Bo Bice. All three had new albums to publicize. Bucky said, "Being on *American Idol* for me was probably the most intense thing and smartest thing I ever did. I think it's safe to say now, that you don't have to win the show now to go on and do something with yourself." Since being on the show, Bucky had recorded an album, released two singles, and filmed a music video. Similarly, Bo Bice was grateful to *American Idol*: "There's no doubt in my mind that people would have no idea who Bo Bice is without *American Idol*. I'm just proud to be alumni as can be. Thank you *American Idol*" (April 2, 2008). Their successes demonstrated that *American Idol* can provide the ending of the Horatio Alger narrative, even if a contestant isn't the overall winner. The only elements the contestants needed to start their rise to fame and fortune were their humble backgrounds.

Continuing the theme of checking on past contestants, the show demonstrated another avenue of success that some contestants embraced: Broadway. Ryan introduced a package highlighting the success of Tamyra Gray and Clay Aiken by noting that their names were "up in lights," another version of fame. Tamyra commented on her fame saying, "I still get recognized as Tamyra Gray from *American Idol*. I love my life right now" (April 23, 2008).

The next week Ryan pointed out a couple of successful former *American Idol* contestants sitting in the studio audience, saying, "This season we're all about giving props to our idol alumni. Wanna draw your attention to our very own Gina Glockson and Constantine Maroulis, doing *American Idol Extra*" (April 30, 2008). Gina and Constantine hosted an ancillary *American Idol* television show that supplemented the primary show with gossip and extra performances. However, sitting directly in front of Gina and Constantine was another former contestant, Ace Young. Ryan did not mention him. The apex of Ace's career seems to be his time as a contestant on *American Idol*. Thus, Ace represented those contestants for whom *American Idol* was the peak of their careers and the pinnacle of their Horatio Alger narrative.

CONCLUSION

For the contestants, success is achieving their dream. The dream is to achieve the Horatio Alger, rags-to-riches journey. This is a dream not just of becoming a professional singer, but also of becoming a celebrity through exposure on the show. In order to fully comply with the Horatio Alger narrative, humble beginnings, or at least a convincing story of humble beginnings, are necessary in order to rise above them. Not a single contestant claimed an upper-class lifestyle before the show. Every contestant is made to seem as if he or she comes from a modest, unassuming background. For example, even sixth-season winner Jordin Sparks's family circumstances were downplayed—her father was a professional football player in the NFL. The contestants follow this lower-class requirement, no matter where in the world they come from, hoping to overcome their initial position in life through talent and exposure. Moreover, the immigrant discourse envisioned an ideally equal America, free from discrimination and intolerance. Without these issues to overcome, everyone, including immigrants, has an identical opportunity to realize the rags-to-riches narrative.

The *American Idol* discourse supported, enhanced, and sustained an idealized rags-to-riches American mythology, which in turn aids in maintaining the dominant acceptance of economic, governmental, and national ideology. Belief in the American dream, which *American Idol* demonstrates is worthwhile and certainly achievable, buttresses the extant social paradigm.

Chapter 4

Contestant Archetypes

Archetypes contain common characteristics found across cultures and throughout time. The discourse analysis of the first seven seasons of *American Idol* revealed several visible archetypal characters. Each archetype is identifiable and comes with a host of expected traits and behaviors. Observable archetypes include the girl or guy next door, the rocker guy and girl, the R and B diva, the male and female country singers, and the blue-eyed soul singer, among others. On the program, contestants adhered to an archetype identifiable through similarity to historical Americans. Conformity to these archetypal figures provides a simplistic way of identifying contestants, relegating each to a preconceived perception. As archetypes naturalize the status quo, akin to myths, the *American Idol* archetypes function to espouse cultural race, gender, and sexuality norms.

An archetype provides a means for a personal connection to the program for each audience member, as he or she may recognize something of him or herself in an archetype. Moreover, each archetype may represent a portion of the general *American Idol* audience. *American Idol* archetypes also may resonate with archetypes found throughout many cultures and epochs. Each season, the majority of the top twelve contestants correspond with certain determined archetypes. In the earlier seasons of the show, the contestants' archetypes were fleshed out through interviews, backstage access, and performances. Later seasons relied more heavily on heuristics, linking contestants to previously established archetypes. However, season seven marked a return to a focus on the contestants with more backstage access and interviews, and less of an emphasis on guest performers and mentors.

ARCHETYPES

According to Lule (2001, p. 15), archetypes consist of the original framework, "patterns, images, motifs, and characters taken from and shaped by the shared experience of human life, that have helped structure and shape stories across cultures and eras" These archetypes could be mothers, heroes, villains, tricksters, or others and "help create fundamental, archetype stories" by being "at the heart of human storytelling" (p. 15). They offer models for audiences to understand life and guide action (Lule, 2001). These mythic characters are "patterns and models, born from human experience, to imitate and adapt" (p. 30). Thus, archetypes usually strongly exemplify one aspect of human nature and consistently recur through time. Certain archetypal stories and archetypal figures are found in multiple cultures, indicating that the universal human experience can produce similar ideals (Lule, 2001).

A stereotype is a concept similar to an archetype. The term stereotype was first circulated by Walter Lippmann, in his book *Public Opinion*, published in 1922. A stereotype is a representation of people, created "by endowing them with particular recognizable characteristics which are implicitly or explicitly attributed in general to the social group of which they are a member" (Marris & Thornham, 2000, p. 198). The "collection of signs that make up the stereotype mobilize a social mythology about the group" and is usually considered negative (p. 198). Alternately, a stereotype could be considered a conscious social convention used to identify a person (Fiske, 2000).

Dyer (2000) noted that stereotypes are used as an ordering process, a short cut, used to refer to "the world" and express values and beliefs. Dyer expressed that "it is not stereotypes, as an aspect of human thought and representation, that are wrong, but who controls and defines them, [and] what interest they serve" (p. 246). Thus, people use stereotypes to quickly draw understanding based on circumstantial evidence. Stereotypes may be defined as models with at least one exaggerated characteristic and are usually considered either extremely positive or extremely negative. Archetypes often resound through many different cultures and time periods, whereas stereotypes are usually native to a particular time and location. When studying archetypes or stereotypes, researchers typically look for dominant attributes that are used to define a person or social group. Both archetypes and stereotypes exist within the program *American Idol*. This study examines archetypes, as many of the *American Idol* characters intertextually refer to previous archetypal incarnations, transcending the time period in which the show aired. For example, season-seven winner David Cook, a member of the rock and roll archetype, contained elements of previous rock and roll figures from different time periods, such as Elvis Presley or Bruce Springsteen. At the same time, there were stereotypical aspects that remain bound in the 2000–2010 decade, such as certain sartorial features.

Archetypes and *American Idol*

While from year to year an *American Idol* archetype may alter somewhat, in general there are more similarities than differences between various seasons' archetypes. Typically, the show itself labels the contestants' archetypes within the show's discourse, for example, describing a contestant as a rocker or girl next door. Indeed, contestants may want to fit an archetypal role, yet differentiate themselves a little from their predecessors. Minor differences aside, fitting into and exemplifying an archetypal character helps to perpetuate associated ideals and standards.

At the season-seven Philadelphia audition, which aired January 15, 2008, an unusual young woman auditioned. She discussed her love of the *Star Wars* films and had arranged her hair in ear-covering Princess Leia buns. Despite her unconventional appearance, she seriously auditioned believing that her voice warranted the valued yellow ticket to Hollywood. After being informed she was not good enough to continue, she railed against the show. She ranted that the producers wanted a "typical, superficial girl with the hair and pretty make-up" and that all such women were an "imitation of something else, not different." She continued that the show "needs more diversity" and that they choose all the "same normal, pop people." But as she spoke, the images being shown on television were of all sorts of people emerging from the audition room: people of all ages, races, sizes, levels of beauty, and dressed in a variety of different styles. They were not all the same, but rather were all individuals with a supposed equal chance at winning. The show's discourse asserted that the contestants who succeed not only have great talent and authenticity, but also are individually unique. However, the discourse analysis of the first seven seasons found that while the people may be unique, several archetypes appear repeatedly with slight variations. These archetypes illustrate a duality in the *American Idol* discourse: an emphasis on conforming to an established image while simultaneously stressing the importance of "being yourself." Contestants are encouraged to "be themselves;" as long this corresponds to a preconceived notion of archetypal identity. Thus, the importance of ostensible individuality remains despite the presence of conventional character models.

American Idol Archetype Variations

Contestant archetypes are recognizable through musical genre preference, physical appearance, and spoken identifying discourse. Each contestant is labeled repetitively through discourse as an archetype. This helps the contestants to be easily identified and understood, and also helps to differentiate contestants within the same season from each other. These archetypes are thus created through the show's labeling and editing of scenes, as well as choices the contestants make in what to sing, how to dress, and verbal utterances. As an example, all of the ultimate *American Idol* champions were easy to distinguish through archetype. The first winner, Kelly Clarkson, was the quintessential girl-next-door pop singer. The second winner, Ruben Studdard, was the essential rhythm-and-blues guy, while third-season victor Fantasia was his attractive and thin female counterpart. Fourth-season winner Taylor Hicks exemplified the blue-eyed soul archetype. Fifth-season winner Carrie Underwood was a female country singer, while sixth-season winner Jordin Sparks was another girl next door. Seventh-season winner David Cook was the emotional rocker. Each archetype picks up aspects of a contestant's personality and amplifies these, creating variations within each archetype.

Girl Next Door

Kelly Clarkson, the first *American Idol* winner, originated the girl-next-door archetype on the show. The "girl next door" is friendly and nice, makes mistakes, and is likable. Mary Ann from the program *Gilligan's Island* or Judy Garland in *Meet Me in St. Louis* or as Dorothy Gale in *The Wizard of Oz* could be considered her non-*American Idol* predecessors. Other "girls next door" in subsequent seasons included Kimberly Caldwell, Julia DeMato, Diana DeGarmo, Lindsey Cardinale, Melissa McGhee, and Jordin Sparks. While most of these women are Caucasian, a category of Asian "girls next door" exists as well and is discussed later in this chapter. Season-six winner and girl-next-door Jordin Sparks has a mixed-race heritage that she never discussed on air. This omission, Jordin's light skin color, and the lack of other African American "girls next door" suggest that African American women were not or could not be considered "girls next door." This may be related to the long history of American racial segregation and discourse of exoticness concerning the female African American. The lack of African American "girls next door" points to the Caucasion viewpoint of the program, the contestants, and the audience.

The first "girl next door," Kelly Clarkson, was from the small town of Burleson, Texas, and was working as a waitress when her friends convinced her to audition for the unknown show. Throughout her season, again and again she was called a "small-town girl," "normal," and a "girl next door." She herself noted that she was "just a small country girl with a great opportunity, that's what I am" (June 19, 2002).

First-season cohost Brian Dunkleman explained that "making the big time hasn't changed the pride of tiny Burleson, Texas." Randy complimented her by saying she had "down home, small-town girl style, but still classiness to her." Even Simon said, "You're a nice person" (July 31, 2002). Kelly was described as nice, classy yet with a small-town demeanor and manners, and unchanged despite singing on national television. Subsequent "girls next door" were also depicted as hailing from small towns.

Kelly described herself : "I am a what-you-see-is-what-you-get girl." And when asked about her childhood goals, she said that "I had dreamed since I was a little girl that I wanted to be on the Grammys or some award show and sing on there, if that could happen that'd be so cool." Simon pointed out the negative aspect to being the average "girl next door," calling Kelly "unmemorable" (June 19, 2002).

On Kelly's visit home to Texas, she filmed commercial spots to air on local television. Repeatedly, she botched her lines introducing herself. She exclaimed, "I'm Kelly Clarkson, from Burleson, Texas. I just said that wrong!" She was possible to relate to due to her imperfections and mistakes. She was an average person, made blunders, and became confused—her flaws made her one of the audience. Brian Dunkleman commented on her mistakes, "True to her style, she handled it the Kelly Clarkson way" (August 28, 2002). The Kelly Clarkson way, it seemed, was the normal way people make mistakes and then laugh at themselves. Kelly demonstrated good humor and a down-to-earth nature.

Her unpretentious and earnest nature was also exemplified in her offstage interviews and reactions to not being eliminated. For example, in one memorable interview she faced the camera with no make-up, her hair in messy pigtails, and stumbled over her words. She appeared disheveled, unpolished, and came across more like a friend at a sleepover than a pop star. Accordingly, later, she gasped in surprise when informed she had made the top ten.

And just like a typical "girl next door," Kelly came from a strong family. Even if a girl-next-door's family was not the nuclear ideal, the strength of the family's love and bonds were always emphasized. With a few notable exceptions (such as families of contestants Syesha Mercado and Kellie Pickler) dysfunctional families were rarely discussed on *American Idol*. On a final episode, when Kelly visited her hometown, a family member said, "To us, she'll always be Kelly." No matter how famous Kelly was or could be, to her family she remained the same average girl. This also brings Kelly as celebrity down to the level of an average person. Later in the show Paula noted to Kelly, "Your family is so proud of you" (August 27, 2002). The discourse of show, while always emphasizing family, placed Kelly as a girl-next-door celebrity.

Another attribute that the show discourse ascribed to Kelly was humility. On the finale, Ryan asked Kelly how she felt about being on the show for the past couple months. She answered, "We're kind of tired, we love singing" (September 4, 2002). She humbly answered for everyone, not just herself. She was so modest, she almost didn't want to take credit for her success. At the same time, her hard-luck story of cocktail waitressing placed her as working class, ripe for the rags-to-riches journey.

Kelly's naïve nature and average physical appearance, combined with her spectacular vocal talent, facilitated her origination of the *American Idol* girl-next-door archetype. Additionally, her choice of songs helped fashion the archetype. Kelly chose popular songs that were mainstream, recognizable, and often played on the radio. Her genre of choice, pop, cemented her status as first *American Idol* girl-next-door pop singer.

Other Girls Next Door

After Kelly, other contestants shaped their own version of the girl next door. However, while they may have been somewhat different than Kelly, all of these women were gregarious and claimed to prefer pop music to other genres. Jasmine Trias and Ramiele Malubay were the Asian girls next door, emphasizing likability as well as exoticness of being Asian. Jessica Sierra, Katherine McPhee, and Haley Scarnato were the sexy girls next door, flashing cleavage or leg, often flirting with Ryan or the judges. There were two funny girls next door, Amy Adams and Mikalah Gordon. Additionally, there was the pop-meets-rock girl next door, Carly Smithson. Carly professed to a love of pop music and usually sang pop, but her physical appearance was decidedly rock, particularly her tattooed arm. Other women may have attempted to fit this archetype before the top twelve contestants were announced but did not have the chance to fully develop their personas through discourse and screen time.

Jasmine Trias: Asian Girl Next Door

Seventeen-year-old, third-season contestant Jasmine Trias self-identified as Filipina and Hawaiian. Jasmine was the first Asian person to make the top twelve on *American Idol*. Her parents emigrated from the Philippines to Hawaii, where Jasmine was born. The Hawaiian and Philippine audiences offered her support in the form of votes, signs, and screaming applause. Jasmine consistently looked cute, not sexy, and always sang pop songs. Discourse about and by her throughout the season emphasized her American qualities, her small-town background, her love of Hawaiian culture, and her close-knit family. She earnestly spoke in a package of scenes from her Hawaiian home town:

> I live in a small town called Mililani on the island of Oahu. I live with my mom, my dad, my brother, and my little sister. This is my house (points to house). Come check it out! And it's so good to be home (flopping onto bottom of bunk bed). March 17, 2004

This package pointed out that although Jasmine may come from an exotic vacation place and appear physically different from the majority of the contestants, her upbringing was the same as any American's. Her family's house was a suburban split-level, the type found across the country. In it, she had to share a room with a sibling. The following week, she further explained her background, "I'm Filipino, Spanish, Chinese. I ended up here. Well, my parents are from the Philippines and I was born here" (March 23, 2004). Her heritage may be a mix from a variety of cultures, but she stressed her American identity. Her friendly nature, non-threatening and non-sexy demeanor, affinity for pop music, and earnest and likable tone corresponded with the girl-next-door archetype. She was the Asian girl next door, a petite and somewhat exotic version of Kelly Clarkson.

The next Asian girl next door was Floridian Ramiele Malubay. The season-seven contestant was linked to Jasmine Trias from the moment she appeared on the show. Ramiele auditioned on the January 30, 2008, episode. She said before her audition, "I want to be the first Asian-American idol. I know Jasmine Trias, she's Filipina also. She got pretty far and if she can do it, I can do it." The show then jumped from the interview with Ramiele to a package of highlights from Jasmine's time on *American Idol* and her subsequent career successes. This directly linked Jasmine and Ramiele before Ramiele even sang on air. Indeed, the show went on to portray Ramiele similarly to how Jasmine had appeared: friendly, caring, and with a love for pop music. Ramiele was a little spunkier and more hip than Jasmine, but retained the same earnest girl-next-door characteristics, with a hint of exotic due to her embrace of her Filipina heritage.

That Asian girls-next-door existed and African American girls-next-door did not speaks to the long history of American racism toward African Americans and the stereotyping of Asians as the "model minority." Asians are stereotypically viewed as intelligent, industrious, and non-threatening and are thus more acceptable as a typical neighbor. African American men are frequently portrayed in media as criminals (Chiricos and Eschholz, 2002) and women are historically viewed as promiscuous (Magubane, 2001), seemingly precluding inclusion on *American Idol* as good neighbors.

Sexy Girls Next Door

Sexy girls next door included Katherine McPhee, Jessica Sierra, and Haley Scarnato. Jessica and Haley, in particular, were always ready with a friendly smile, a low-cut blouse, and a suggestive comment. Katherine, on the other hand, wasn't painted as overtly sexy, but still retained sex appeal through beauty and dress. These girls were sexy, but still fit the girl-next-door archetype through music genre preference and a primary quality of being nice.

Katherine, referred to as Kat, was an aspiring actress from L.A. Pop music was her preferred genre. Her sex appeal was obvious through her glamorous, low-cut dresses, while her seemingly sincerely kind nature made her fit the girl-next-door archetype. She was always nice to everyone, always smiling, and was a beautiful, sexy, and normal girl next door.

Jessica and Haley, in contrast, were blatantly sexy both through their dress and their flirtatious nature. They made suggestive comments and showed a lot of skin, but still managed to remain wholesome through their eager, toothy smiles and home-life packages. These contestants played up their sex appeal through costuming, and the show embraced it as well. In a telling example, on the season-six top-eight performance night Haley sang a pop song and danced suggestively. She wore short shorts and high heels. Simon judged her by saying:

> I think you have a very good tactic at the moment, Haley—wear as least amount of clothes as possible. (The camera panned slowly up her body, from feet to head.) Because look, I'll be honest with you—you can't do [sic] in the competition based on your voice, because there are much better singers. All you can do is have fun. (April 10, 2007)

Simon assumed that Haley's best assets were her figure and sex appeal. Akin to the Mary Ann character on the television show *Gilligan's Island*, Haley was the wholesome, yet sexy, girl next door. She was wholesome through her innocent grin and overall congenial demeanor; she also fit a stereotypically sexual appearance.

Female versions of next-door neighbors were allowed to be sexualized, albeit within a wholesome frame; the next-door men were never depicted as sexualized. Sexualized women are nonthreatening and feature predominantly in media; sexualized men, while increasingly visible in general media, are typically absent from family-friendly TV fare. Thus, children are exposed to sexualized women early on, while sexualized men are omitted.

Pop Meets Rock

Season-seven contestant Carly Smithson looked very rock and roll with her arm of tattoos. Midway through the season she even had the Roman numeral

VII tattooed on her knuckles in homage to her season. However, her past was decidedly pop, as was her musical genre of choice. She was the girl next door meets rocker chick.

Carly noted that she owned a tattoo shop with her husband in San Diego and also worked as a bartender. Her husband had tattoos on his face and was rarely shown on camera. His lack of screen time could have been due to his unconventional appearance possibly contradicting her girl-next-door archetype, or perhaps someone (an editor or producer) thought his appearance would upset at-home viewers, prompting a channel switch. Her husband's alternative appearance aside, and despite Carly's two jobs, she insisted that,

> I'm definitely a homey person. I love my house. I love to clean, I love to cook. With all that there's probably nowhere else that I feel as home as on the stage. As much as I'm missing home, I'll give it up to be on *American Idol*, 'cause this is really what I want to do. (February 27, 2008)

She made herself out to practically be a housewife. Despite her tendency toward being a homebody, she always wanted to sing. Along those lines, she noted that, "I always wanted to be a pop star when I grew up. I always wanted to be Madonna or you know, Kylie Minogue or those kind of people" (March 25, 2008). While Madonna and Kylie Minogue may not be girls next door, they are pop stars. And no matter how rock or tough Carly physically appeared, her sweet nature, desire to be a pop star, and consistent choice of pop music songs were consistent with the girl-next-door archetype. Carly looked rock, but identified as pop and sang pop music. She was the intersection between pop and rock.

Guy Next Door

The male counterpart to the girl-next-door archetype is the guy next door. Just as the prototype for the girl next door existed in season one, so did the guy next door: Justin Guarini. Justin was the quintessential *American Idol* guy next door. He was attractive and sang pop music. His primary characteristics were his kind nature and goofiness. The character of Dobie Gillis, from the early 1960s sitcom *The Many Loves of Dobie Gillis* may be considered a guy-next-door predecessor. Over the next seasons, several contestants fit the guy-next-door archetype, including Jon Peter Lewis, Matthew Rogers, Chris Sligh, Sanjaya Malakar, David Archuleta, and Michael Johns. The guy-next-door archetype also has a nuanced subcategory, the nerd next door, typified by Clay Aiken.

The season-one discourse positioned Justin in the same way as Kelly. Justin was nice, friendly, attractive, silly, and sang pop songs. His appearance was non-threatening and his affable nature was appealing.

Prior to auditioning for *American Idol*, the show described Justin as a door-to-door salesman. While employed as a door-to-door salesman, a working-class job, Justin was also an aspiring musician. On the top-five performance show, Justin explained, "on the stage is where I can be me, and just be happy, and just let it all go" (August 13, 2002). His love of singing and performing was situated as natural. Justin himself was always described as normal and nice. Throughout the season, Justin was referred to as "the natural." This label implied that Justin being himself was equal to him being a performer.

Justin's physical appearance was cute and appealing, hardly sexy or threatening. While the show repeatedly called him a "heartthrob" or similar labels, he was goofy and nice, not sexual. He wore bell-bottom jeans and tank-tops and he had an extremely curly Afro and honey-colored skin. Aesthetically, his facial features were very symmetrical and pleasing, and he could have been sexy; however, his friendly nature belied any attempt to be sexy.

Like Kelly, Justin came from a small town: Doylestown, Pennsylvania. Despite being in L.A. and performing on national TV, Justin said he never stopped feeling "like a small-town kid" (August 28, 2002). On the season finale he said, "I can't believe it, I still can't believe it" and that "it's been a crazy journey from sitting on the cold hard pavement in New York in that line to sitting in the Kodak Theatre" (September 4, 2002). He was sometimes humble, but always confident. Accordingly, he also said, "this is a gift that I've been given, and I get to share that gift with whoever's listening and that makes me feel good" (September 4, 2002). While claiming to not forget his small-town roots, he was self-assured and certain of his talent.

In addition to Justin's unassuming nature, he also valued his fans. On the finale after Kelly won, Justin thanked his supporters when given the chance to speak:

> I am so blessed to be up here; it's all because of you people (gestures to audience). I'd like to thank my family, the backbone of my support, all the fans everywhere, the judges, I've learned so much from you guys. Simon, thank you for being so honest with me. Paula, you've been my heart throughout this whole competition. (September 4, 2002)

In another parallel to Kelly, Justin had a strong family background. During the week the top-three finalists competed, his family members were introduced and expressed their support of him. Justin said, "I just can't wait to come back and see everyone" and seemed to enjoy spending time with his family (August 27, 2002).

The show discourse effectively positioned Justin as Caucasian by not discussing any alternative, just as it later did with Jordin Sparks. His family at home was Caucasian; his biological father, who rarely appeared on the show, was African American. Justin's thoughts on his heritage were never discussed on air. Regardless of his unclear relationship with his biological father, his family background appeared strong and average. Generally, on television, Caucasian is the fallback norm, as there is a lack of representation of other races. Justin's African American heritage would have complicated the guy-next-door archetype. Similarly, while guy-next-door David Archuleta occasionally mentioned his Honduran heritage, this was overshadowed by emphasis on other personality traits, such as likability and humility. All guys next door were positioned as Caucasian by omission of any other racial discussion.

Nerd Next Door

While Justin Guarani and others like him fulfilled the guy-next-door niche, another version of the guy next door existed as well: the nerd next door. The nerd next door was similar to the guy next door, only less cute and geekier. Like the guy next door, he sang pop music songs. He was approachable, and may have altered his image to be more attractive, but never lost the initial nerd impression that he started the show with. This subset is even less intimidating than the guy next door and often has young teen and pre-teen female fans. The nerd next door is a stereotype, related to the contemporary time period, while the guy-next-door archetype transcends time periods.

Television nerd stereotypes appear to have become visible in the late 1970s, with "The Nerds" sketch on *Saturday Night Live*, starring Gilda Radner and Bill Murray. Later, nerds appear in the 1980s sitcom *Head of the Class*, and memorably appeared throughout the 80s and 90s as Screech Powers on *Saved by the Bell*, Paul Pfeiffer on *The Wonder Years*, Brian Krakow on *My So-Called Life*, Steve Urkel from *Family Matters*, and as the title character in *Doogie Howser, MD*. Over the years, *American Idol* nerds next door have included Clay Aiken, John Stevens, and Anthony Federov. The first *American Idol* nerd next door was Clay Aiken, who typifies this sterotypical subset of the guy-next-door archetype.

Clay, like many of the subsequent nerds next door, was exceedingly thin, had an unstylish hair cut, and wore glasses. Moreover, at the start of the season Clay dressed as a nerd and not as a pop star. While Clay eventually altered his image into more of a pop star than a nerd, he couldn't shake his initial geek impression. Through the end of the second season, Simon continued to put down Clay's appearance, even though it had somewhat changed mostly due to sartorial revisions. Many of the later geeky contestants found this to be true for them as well, particularly Anthony Federov. Simon's comments confirmed the authenticity of the geek image—these contestants truly were authentic nerds—and not manufactured to seem that way.

Throughout the discourse of season two (and often in subsequent seasons) Clay was repeatedly referred to implicitly as a nerd and overtly as not attractive. He was repeatedly told he did not look like the American idol. However, as Simon said, "Clay, you don't look like a typical pop star. But you are unique" (March 4, 2003). While Clay was physically not the typical pop star, he was different than anyone else.

Before auditioning for *American Idol*, Clay worked as a special education teacher. During the Hollywood round, Clay said he felt compassion for the mentally handicapped and declared that if he were to win, he would advocate for mentally disabled people. While Clay was earnest and sincere, this charity was not very sexy or appealing, and only enhanced Clay's geeky image. Of course, a desire to help those in need is admirable, but not a trait typically displayed by contestants on *American Idol*, with the exception of the *Idol Gives Back* episodes. While this avowal may have endeared Clay to audience members, it did not make him seem cooler or more hip.

In line with the guy-next-door archetype, Clay also sang pop music songs. He often sang ballads, as well as many instantly recognizable hits by the likes of Journey, Elton John, The Four Tops, Billy Joel, Edwin McCain, The Righteous Brothers, Bobby Darrin, and Simon and Garfunkel. Every song he sang could be considered pop music.

The overall discourse of the show, including verbal comments, Clay's dress, and pop song choices made Clay the quintessential nerd next door. Notably, the guy and geeks next door were not positioned as blue collar or as homosexual. Everyone on the show was presumed to be heterosexual; no one claimed any other label while on the show, even as rumors swirled in the press. This complies with mainstream media, which perpetuates the heteronormative status quo.

Other Nerds Next Door

Other nerds next door included Anthony Federov and John Stevens. Both were very thin and pale, and dressed unfashionably. Each performed pop music. Moreover, the show discourse situated them neatly into the nerd-next-door stereotype.

On the night that the top-eight men performed in season four, Simon commented after Anthony's performance of a Marc Anthony song:

> The problem I have, Anthony, is that you're a nice guy. You remind me a bit of Clay. You know, you're sweet. But now, you're singing Latin, Spanish, Marc Anthony. To me I found it all a bit odd. You have as much Latin flair as a polar bear. You're nice. (March 7, 2005)

Likewise, season-three contestant John Stevens was also a nerd next door. John was sixteen years old, and had bright red hair. From his initial audition, John professed his love of Frank Sinatra, Dean Martin, and the rat pack. Loving a bygone style of music adored by geriatrics did not up his cool quotient, nor did always wearing a suit. John addressed his appearance saying, "I will be making some alterations in my image, but I will be keeping the magic suit while becoming more pop-ish in my song selection. This is what I want. I'm going for it" (March 2, 2004). John recognized that he matched the nerd-next-door image and wanted to have more of a pop-star image. He vowed to retain his lucky suit, which was a nerdy costume for a sixteen year old to continually wear, but tried to move away from the crooners' songs. While pop songs are usually a component of the guy-next-door image, John's geeky image was greatly influenced by his penchant for songs from the 1950s and 1960s. John did try to alter his image by singing more pop songs, but his crooning singing style and image proved difficult to change. In a third-season upset, John finished in sixth place, remaining after Jennifer Hudson was eliminated.

Female Rocker

Another consistent archetype is the rocker. In the first season, the show discourse identified two female rockers, Ryan Starr and Nikki McKibbin. Both professed to love rock music, dressed as rockers through stereotypical clothing, and were noted as blue collar. Subsequent female rockers included Vanessa Olivarez, Gina Glockson, and Amanda Overmeyer. All of these women are Caucasian. Female rockers from the past include Janis Joplin and Joan Jett. This archetype includes elements of ostensible rebellion; this recalls ancestral antecedents of women such as Joan of Arc. A long tradition of non-conformists supports the rock archetype.

The season-one discourse asserted that both Ryan and Nikki were rockers. The two had similarities: both were attractive and thin, wore outlandish rock clothes, and were repeatedly called rock and not pop. While Ryan was labeled the "resident rebel" (July 30, 2002), Nikki was also tagged as "the rocker" among the top five contestants (August 13, 2002). Nikki finished the season in third place and subsequently received more air time than Ryan, who finished in seventh place. Possibly because Nikki was featured on the show more than Ryan, more qualities attributed to Nikki compose the female rocker archetype and appeared in subsequent reiterations of the original.

In packages about her home life and in interviews, the show's discourse stressed Nikki's blue-collar, working-class background. As a single mother, she struggled financially working as an employee of a karaoke company. She always sang but never earned enough of a living to be a professional singer. This blue-collar background came to be a significant characteristic for those contestants corresponding to the rocker archetype.

Additionally, according to her family and herself, Nikki was a "wild child," a true rocker who never lived by the rules (August 27, 2002). She was a nonconformist. The show discourse repeatedly mentioned that "Nikki is such an individual" (August 20, 2002). Upon Nikki's elimination, Paula predicted her "originality will propel you to superstardom" (August 28, 2002). She was unique and distinctive, yet familiar as an archetype to audience members. Nikki the rocker emulated previous radicals along the lines of Janis Joplin and Courtney Love.

Although Nikki may have been an individual, she still had the support of her family, her mother and her son, and her hometown. She claimed that "no matter how famous I am, I'm always, always, going to come back home" (August 27, 2002). Despite her rocker orientation and desire to flout societal conventions, she still loved her family and valued her roots, a common denominator that may have helped the audience to identify and empathize with her.

Nikki also sang songs with a rock edge. Artists she covered included Melissa Ethridge, Alanis Morissette, Pat Benatar, Stevie Nicks, and Janis Joplin. When Nikki sang a song not known as rock, she endeavored to make it seem more rock through her vocal performance and the song arrangement.

Other Female Rockers

Throughout the ensuing seasons of *American Idol*, other various versions of rock women appeared, including contestants Gina Glockson, Vanessa Olivarez, and Amanda Overmeyer. Fourth-season contestant Nadia Turner attempted to create her own archetype of rock-meets-funk, but instead of fusing the two, she never fit any archetype at all.

On season six, however, Gina Glockson clearly embodied the rock archetype as evidenced by her appearance, musical preference, and the show discourse. On the night that the top nine contestants performed, Gina wore heavy eye makeup, fingerless gloves, boots, a chain belt, and several necklaces. She also had pink streaks in her brunette hair and showed off her pierced tongue. After she sang, Randy commented, "That was a nice performance from the rocker girl" (April 3, 2007). However, she was eliminated the next night and finished her season in ninth place. During the rock-themed week later in the season Gina sat in the theatre audience. Ryan Seacrest commented to her "it's rock week, it must be killing you" (May 1, 2007). Even though she was no longer competing on the show, she still retained the mantle of rocker.

The smart rocker

The following season, contestant Amanda Overmeyer also strongly fit the female rocker archetype with her raspy voice, her bleached-blond hair with black streaks, and her love for motorcycles. While she fit the rocker archetype, Amanda tweaked it a little, differentiating herself from her predecessors: she was portrayed as smart. Although she came from a rural area, wore rock clothing, and loved her Harley motorcycle, Amanda was a nurse, which placed her squarely in the middle socioeconomic class. She was not blue-collar, which differentiated her again from her predecessors. At her audition in Omaha, Nebraska, Amanda wore leather pants, heavy makeup, and a cross on a choker, and sang a Janis Joplin song. The judges loved her. Randy said, "The rock-and-roll nurse is through to Hollywood." Paula later compared her by saying "She's the female Chris Daughtry [an earlier male rock contestant]" (January 29, 2008). From her first moment on the show Amanda clearly matched the rock archetype through discourse, appearance, and musical genre preference, even as she altered it by class membership.

The verbal statements assigning Amanda to the female rock archetype continued each time she appeared on the show. During Hollywood week, "Amanda rocked her way straight through to the next round" (February 12, 2008). The next week, Amanda herself explained: "People have embraced me as being rock. My goal when I go out there is just to solidify the presence of rock and roll, I guess" (February 20, 2008). Accordingly, her outfits included tight jeans, cleavage-showing tank tops, big cross necklaces, lots of bracelets, large belt buckles, heavy makeup, scarves, and hair that was a frightening combination of bleached blond and extreme black.

While Amanda embraced most rock characteristics, she altered the archetype a little through her clear intelligence. While the previous female rockers may have been bright, Amanda ably demonstrated her intelligence through articulate responses, reading on camera, and through her own examples. For instance, she explained:

> Most people already know that I ride a Harley and I'm a nurse. Another thing people may not know is I tend to be kind of a bookworm. I like to read quite a bit. I like reading biographies of rock icons that I like. I'm one of those 'knowledge is power' people and you can't ever be too over-educated. And I'll be the first one to recognize what I'm not. (February 27, 2008)

She also noted "reading to me is just a really good form of entertainment." Later she clarified that she was "just reading a music business book, getting myself educated on what may come" (February 27, 2008). Of all the contestants in all the seasons, Amanda was the only one actively shown researching how to prolong and advance her career.

Most of Amanda's attributes fit the female rock archetype. Her intelligence and class merely shifted the archetype a little, making it her own particular version. The rock archetype was clearly identifiable even as it modified.

Male Rock Archetype

The corollary to the female rocker is the male rock archetype. Unlike the female archetype, the male archetype did not surface until season four, when Bo Bice and Constantine Maroulis competed. While Constantine and Bo competed in the same season, Bo was the first contestant to exemplify the *American Idol* male rock archetype, which was constructed through discourse, physical appearance, and song selection. Bo's archetype was more clearly defined than Constantine's, whose pop song selections muddled his archetypal identity. Subsequent incarnations of the rock archetype modified a little. Season-five competitor Chris Daughtry was the rocker father, while season-seven competitor David Cook was the emo rocker. Qualifying characteristics for these rockers were a working-class background and the common element of struggling to succeed in the music industry, typically as a band front-man. Additionally, all of these men are Caucasian.

From the moment Bo Bice auditioned, it was clear he was a rock singer. He auditioned with The Allman Brothers Band song "Whipping Post," a southern rock anthem, which was shown on the night the top three finalists were revealed, May 11, 2005. After wailing through a section of it, he spoke with the judges. Bo said, "I think *American Idol* could really use a kick of some rock, some blues, some soul. One thing we haven't really seen is somebody that gets up there and puts out the vibe of a rock star." Bo was the rock star.

Early on, Bo discussed his job prior to being on the show. He explained:

> They told me I could choose, *American Idol* or my job, so I said, well, I'm gonna have to say *American Idol*. They said, well, you need to quit. And I said, I'm not gonna quit. If I quit I don't get unemployment. So, they fired me. And I got unemployment for a couple months. (February 28, 2005)

Being unemployed illustrated the lengths he would go to achieve his dream.

Bo discussed playing with his old band prior to *American Idol* as well. He explained that he "chose [the song] 'Spinning Wheel' mainly because my band and I used to play it out around town in Birmingham and throughout the Southeast. Plus, it's got an amazing vocal part that really grabs you." After his performance, Simon complimented him by saying "This stage can either make you or break you tonight. You look like you've been doing it for twenty years" (March 15, 2005). Bo certainly had been struggling in the music industry for several years, if not quite twenty. This struggle helped to legitimize his claim as a rocker. Struggling to receive notice and validation in the music industry is an essential component to the male rock archetype. The agony of dead-end jobs and an unfulfilled dream lend credibility to the emotion of dissatisfaction, which is often found in rock music. Often, discontent and ambition seem to propel rock vocal performances.

Along with illustrating the background seemingly required to achieve success as a rock musician, the discourse on the show repeatedly referred to Bo as a rock singer, further establishing him as an archetypal rocker. Bo himself said it, the judges said it, and Ryan said it as well. On the night the top ten competed, Randy said, "I would love to see a rocker win this year. You're on the right track" (March 29, 2005).

Other verbal discourse spoken by Bo alluded to the rock lifestyle that he lived before even appearing on the show. Bo clearly established the fact that he was always a rocker, on and off screen. Memorably, Bo said that he toured a lot with his band in the 1990s and consequently didn't remember those years well, insinuating he either drank a lot of alcohol or took drugs (March 29, 2005). Later, he said that if he didn't win, he would be fine: "What's the worst thing that's gonna happen to me? I go back home, I play gigs, and I have blast doing what I've been doing my whole life? I'm not worried about it. Dude, I'm good" (April 13, 2005). His dedication and determination belied this attitude.

Bo also originated a rock-star appearance. He had long, shoulder-length hair and a goatee, and usually wore jeans. Bo was from the American south, Alabama, and evoked the feel of a Southern rocker, in the style of rock band Lynyrd Skynyrd, most famous for their song "Freebird," which Bo performed on the top-eight performance night, April 12, 2005. Typically, Bo wore a colorful shirt, often with the cross tattooed on his chest peeking out of the neckline. His accessories included necklaces, bracelets, boots, cowboy hats, leather pants, long duster-style jackets, and sunglasses. Whatever he wore, his appearance screamed rock star.

In addition to his sartorial choices, his song choices and performance style corresponded to his archetype. While Bo sang songs by The Allman Brothers Band, Lynyrd Skynyrd, The Black Crowes, and The Rolling Stones, he also sang songs by Jim Croce, Edwin McCain, and The Temptations in a rock style. He prowled the stage, lifted the microphone stand to emphasize his words, occasionally dropped to one knee, whipped his long hair around, and truly exuded rock charisma. While Bo was performing, the audience could have been at a Bo Bice rock concert instead of in the audience of *American Idol*. Bo Bice was the first male rocker on *American Idol*.

Rocker Dad

The next season, Chris Daughtry said at his audition, "I was inspired by Bo last year 'cause he did a lot of the rock. If he could do it, then maybe I should show them what I have. I have a rock band that I perform with as well" (January 18, 2006). Also like Bo, Chris worked a day job while gigging with his band at night, trying to succeed in rock music. Chris thus filled the rocker archetype. However, Chris's archetype also included a focus on Chris as a father.

Chris's song choices and performances confirmed the rock archetype. Over the course of the season, Chris sang songs made famous by rock bands Bon Jovi, Fuel, Styx, and Creed, as well as songs by Keith Urban, Bryan Adams, and Louis Armstrong in rock style. Chris stalked around the stage, gripped the microphone stand in nearly every performance, and lifted it up to punctuate the lyrics. At times he sang in a forced vocal style akin to rock bands' singers' performance style, pushing the sound out of his throat in a strangled voice.

Moreover, during his appearances onstage and off, Chris looked like a rocker. He usually wore jeans, tight-fitting tops open at the neck, and a chain wallet. He had a shaved head and habitual five o'clock shadow. In addition, he was very muscular and well built. As fitting for a rock star, Chris often wore the color black. He was an attractive rocker who could have seemed a little dark if judged solely on performance. However, the focus on his family life mitigated his rough-edged appearance.

At his initial audition, Chris's wife spoke to the camera, explaining why Chris should receive a golden ticket to Hollywood:

> This is his dream. When he married me, he took on my two kids and he's younger than me. He's only twenty-five and I just felt like, he would've done so much if he was single, but he has a family to take care of. I just want his dream to come true, 'cause he's given me and my kids so much. I'm just so emotional 'cause I know that this is his chance. And I just love him. He's such a good husband and good father. (January 18, 2006)

Statements like this one helped to build Chris's reputation as a good father. Further heightening Chris's good father image was the fact that the children weren't biologically his, making his fatherly devotion even more commendable. He repeatedly spoke about how much he loved his wife and children. In an example, before singing the Johnny Cash song "Walk the Line" Chris said, "The lyrics have a lot of relevance to me, 'cause he wrote it for his wife at the time on how he was going to stay true to her even though he was touring the world" (March 1, 2006). Whenever Chris performed, his wife was shown in the audience, reminding the audience at home of Chris's dedication to his family. Through the discourse about his wife and kids, his rock music preference and style, and discourse about him as a rocker in general, Chris personified the rocker dad archetype.

Emo Rocker

In the seventh season, another version of the rock archetype emerged: the emo rocker. The term *emo* is shorthand for emotional, a contemporary type of popular rock music. Emo music generally has depressing lyrics set to up-tempo rock. Emo rock bands include Fall Out Boy (particularly known for its charismatic bass player Pete Wentz), My Chemical Romance, and Panic at the Disco. Gerard Way, the front man of My Chemical Romance, characterizes the male emo look with his stylish haircut, pale skin, tight black clothing and straight-leg jeans, and proclivity for emotional expression through song lyrics. David Cook embodied this look and style of music, making him the *American Idol* emo rocker. Additionally, his emotional candor was resonant with emo music's characteristics.

Discourse on the show situated David within the rocker archetype. At his audition, he was introduced through a comparison to Chris Daughtry. As an image of Chris flashed on the screen, Ryan's voice-over said, "Thanks to Chris, we met this guy." David then continued the comparison by saying, "I watched Chris Daughtry's audition last night to kind of figure out some sort of a competitive edge, to watch how people who have made it through have done it" (January 29, 2008). For his audition, David sang "Livin' On A Prayer" by Bon Jovi, immediately specifying his rock archetype through musical preference.

Throughout the season, the discourse on the show repeatedly referred to him as rock. As Randy said, describing a performance, "It's like going to another song at a David Cook concert. You got the Eddie Van Halen amp, your Les Paul [guitar], you're throwing picks out into the crowd, you got the voice box solo" (March 18, 2008). A couple weeks later, Randy noted again "You're a rocker but you've been showing that you got this unbelievable range" (April 1, 2008). Similarly, on the night the top five finalists performed, Ryan introduced David by saying, "Now it's time for the rocker's second outing. Say hi to David Cook again" (April 29, 2008).

David Cook, like Bo Bice and Chris Daughtry, struggled for years while playing with bands, attempting to break into the music industry. The show discourse emphasized these credentials, linking him to his predecessors. On an early episode he explained his background:

> For a living, before all this crazy idol stuff started full swing, I was a drink slinger. I bartended at a bar called the Blank Slate, in Tulsa. It was cool because it's also a music venue so I was also able to play shows there from time to time. It was the only job I ever had where you're fully able to be who you are. No one's going to fault you for it; in fact, they're probably going to tip you a little bit better. And I love it, but I certainly would rather be here. (March 11, 2008)

David's time bartending and playing small venues gave him rock credibility through hardship and the struggle to attain his dream.

Moreover, David performed rock music throughout his season, transforming music not considered rock into rock songs, such as songs by The Beatles and Lionel Richie. A new seventh-season rule that permitted contestants to accompany themselves on instruments helped David. Adding an electric guitar on stage upped the rock-star quotient immensely, as well as immediately informed the audience what to expect from the performance. In addition to performing songs that may not have been considered rock, David also performed several rock songs, including songs by Bon Jovi, The Who, and Collective Soul.

On stage performing, David expressed his emotions (further discussed in chapter six) and also looked the part of an emo rocker. He dressed in tight jeans and had fair skin, five o'clock shadow, and a stylishly tousled haircut that was dyed different colors. He often accessorized with ties, vests, jackets, and a chain wallet. His performance style was very emotive and rock as well. Performing with an electric guitar automatically placed him in the rock archetype, but he would also pick up the microphone stand occasionally, reminiscent of Chris and Bo (and Mick Jagger and Steven Tyler), brandishing it to accentuate points in the music. His raspy voice and rock performance style added to his rocker archetype.

The show discourse, David's physical appearance, musical preference, and performance style all worked together to help fashion his version of the rock archetype, the emo rocker. The rocker archetype began with Bo Bice, and continued and was modified by Chris Daughtry and David Cook. All three toiled for years playing music, enjoying themselves but desiring recognition. Their clearly defined archetypes most likely helped them achieve their successes on *American Idol*.

While the rock archetype is clearly defined, there are some crucial differences between the male and female versions. Whereas Bo, Chris, and David were lauded for their experiences and struggles, Amanda and Carly (who blended rock and girl next door) had their previous experiences minimized. The men's pasts were depicted as positive and valuable experiences, preparing them for the *American Idol* stage and further careers. Carly conspicuously had a well-known previous musical career, including a disappointing relationship with a major label. This experience was generally downplayed. When it was mentioned, it was in the context of Carly receiving a second chance, thus positioning her past experience as failure. Concurrently, Carly's history was construed as failure while the male histories were learning experiences.

Moreover, David's intellectual image was promoted throughout the seventh season. In contrast, the judges determined Amanda's demonstration of intelligence didn't fit her rock image. While a rock man was portrayed as smart, in the same season a rock woman was derided for adding this component to her persona. This contrast is discussed further in the following chapter.

Rhythm-and-Blues Diva

Another archetype labeled through the discourse analysis across all seasons was the rhythm-and-blues (R-and-B) diva. The R-and-B divas, all African American, came in many shapes and sizes and can be divided into two different categories through discourse: implicitly sexual, slender divas and overtly sexual, heavy divas. The slender divas were not overtly labeled as sexual through verbal discourse, while the heavier divas were. The show (and the contestants) may have wanted to position the heavier women as sexy instead of fat. Overall, the diva is a sexualized archetype, yet often is also associated with religion. Tamyra Gray and Christina Christian initiated the *American Idol* representation of the archetype in season one, although Tamyra received more air time after Christina was eliminated. Other rhythm-and-blues, sexualized, slender divas include Trenyce, Fantasia, Latoya London, Paris Bennett, and Syesha Mercado. All were African American and often sang rhythm-and-blues songs or gospel music. They were very self-confident and flatteringly dressed, and sang with powerful voices. Rhythm-and-blues music is understood as generally beat-driven music incorporating elements of soul, blues, and African American pop music (Garofalo, 1994). *American Idol* variations of this archetype include the nonsexualized R and B singer, the family-oriented mother, and the overtly sexualized, slender diva.

On season one, the discourse of the show asserted that Tamyra Gray was "the natural" due to her exceptional voice and performance skill (August 13, 2002). In packages, Tamyra was shown singing in church, highlighting her connection with organized religion (August 20, 2002). Subsequent contestants following in her archetype continued this association and often discussed religion, mentioned or thanked God, or were actually shown in church.

Tamyra often sang soul and R-and-B songs, including songs popularized by Diana Ross, Tina Turner, Patti Labelle, and Gladys Knight. Her rendition of "A House Is Not a Home," most familiar to audiences as a Luther Vandross song, garnered praise from Simon as "one of the best performances on television" (August 13, 2002).

Tamyra also dressed in a very sexualized manner, frequently wearing tight pants or jeans and showing her midriff. If not in casual wear, she would wear long-skirted, glamorous dresses, appropriate for a diva. Tamyra was extremely slender and often wore her hair up, accentuating the long lines of her body.

Tamyra's sexual dress, the season one discourse, and her musical preference combined to form an implicitly sexy and slender R-and-B archetype. Moreover, religious connotations and associations with famous rhythm-and-blues divas became crucial additional characteristics of this archetype.

Other Sexualized, Slender R-and-B Divas

Season-two competitor Trenyce was superficially very similar to Tamyra: pretty, thin, and African American, with a preference for the R-and-B musical genre. The show clearly linked the two singers, at one point even showing Tamyra sitting in the *American Idol* audience as Trenyce sang. In a parallel to Tamyra, Trenyce also wore long gowns and sexual outfits. And as Paula noted to Trenyce, "you remind me of a brand new Diana Ross" (March 11, 2003). Trenyce sang songs made famous by Al Green, The Supremes, Whitney Houston, and Chaka Khan.

While Trenyce was similar to Tamyra, other women altered the archetype. Season-three winner Fantasia and season-five competitor Paris Bennett also matched the sexy, slender R-and-B diva archetype, but in different ways. Both embraced humor, adding that element to their archetypes. The discourse about Fantasia focused heavily on family and her daughter, religion, and overcoming hard times. Paris's discourse focused more on having fun, being young (she was seventeen during her season), and being fashionable. Both sang R-and-B songs; neither was very eloquent. Paris added youth, exuberance, and overt sexuality while Fantasia added motherhood, even though she was also a teenager (nineteen years old) during her season.

The show discourse concerning Fantasia repeatedly noted the importance of family, particularly Fantasia's relationship with her young daughter, a toddler at the time of the show. Fantasia, while slim, attractive, and youthful, was positioned through the show discourse as more of a religious mother than a bombshell. Her daughter was frequently mentioned and often shown on television with Fantasia, as was the rest of her family. In fact, on the Barry Manilow-themed performance night, Fantasia sang the song “It’s a Miracle.” She described her version of the song as having more “churchy soul” than his (April 20, 2004). After the judges’ effusive comments, Ryan walked out on stage carrying Zion, Fantasia’s daughter, and Fantasia screamed and said, “This is the miracle!” That one performance coupled the family and religious components that Fantasia added to her sexual R-and-B archetype.

In opposition to Fantasia, Paris Bennett altered the slender R-and-B archetype to contain overt sexuality. While Paris was only seventeen, she was sexualized in a youthful and sometimes explicit way. Other teenage contestants opted for cute over sexy, but Paris’s hip-hop-flavored dance moves contradicted her age. On the March 28, 2006 episode, Paris sang Beyoncé’s upbeat song “Work It Out.” Paris said about her performance, “I get to be fun, I get to be a teen, I get to be seventeen, and I get to dance for all the young ones.” Her dress for the evening was conservative: jeans, a yellow top, a pink blazer, and sneakers. However, while singing lyrics like “I couldn’t wait for the bedroom, so we hit the floor,” Paris bounced all over the stage, danced her heart out, and employed hip-swinging hip-hop dance moves that highlighted her sexuality. Typically, the sexuality of the R-and-B diva was understated, connoted through clothes and body language. Long, formal gowns typically worn by divas promoted a classy version of sexuality. However, Paris brought her sexuality to the forefront in a performance Randy called “fearless, the bomb, and the best of the night so far” and Simon called “precocious.” Paula pointed out “those weren’t kid moves” (March 28, 2006). While Fantasia embodied a family-friendly R-and-B diva, Paris was the young, overtly sexual diva.

Big and Sexualized R-and-B Diva

Another category of the R-and-B diva was the big woman. These women were overweight, embraced their bodies, and were sexual despite not fitting the stereotype of physical female beauty. Like the thin R-and-B divas, the bigger women dressed in a sexual fashion, sang rhythm-and-blues songs, mentioned church often, and were positioned by the show discourse as sexual. Sexualized and big R-and-B divas included Jennifer Hudson, Mandisa, and Lakisha Jones. The primary difference between the thin divas and the heavier divas was the blatant sexuality of the latter.

These women usually wore dresses or tops that showed off their cleavage and curvaceous bodies. They flirted outright with Randy and Simon, using suggestive comments, gestures, and looks. Often, their song lyrics were evocative. An example of season-six contestant Jennifer Hudson’s flirtatious nature occurred on February 10, 2004. Jennifer performed wearing a white trench coat and white, knee-high boots. She asked, "Do you like my outfit, Simon?" Simon, always up for flirting, smiled and replied, "The leather nurse look is always on my list." Jennifer, preening and smiling on stage, rejoined, "Well, don't end up in my hospital, okay?" Simon answered, "I'm feeling sick." As the audience hooted and catcalled, Jennifer said, "I'll take care of you." Exchanges like this added to the teasing sexuality of the big divas. These women were not shy or retiring, but the opposite: brazen, engaging, and playful.

Season-six fourth-place finalist Lakisha Jones was also a big and sexual R-and-B diva. She was a large African American woman who seemed to happily embrace her curves, flaunting her figure through form-fitting outfits, showing cleavage, and wearing sexy boots, high heels, and red lipstick. Simon once described Lakisha as "sassy" (April 3, 2007). After a performance that Simon particularly enjoyed, he said to her "Lakisha, I actually could kiss you after that." Ryan led Lakisha off of the stage to the judges' platform, where Simon stepped down and planted a kiss on Lakisha's lips. He then said, "You are so good" to her (May 1, 2007).

On *American Idol*, heavy African American women can be sexy if they choose to be. Very rarely in film or television are overweight women depicted as sexual. These R-and-B divas owned their sexuality, seemed comfortable with their bodies, and dressed their curves provocatively. They were just as forthrightly sexy, if not more so, than their thin counterparts. They were also more sexualized than their overweight Caucasian counterparts, who fit the girl-next-door or rock archetypes. *American Idol*, while seemingly progressive in the sexualized depiction of overweight African American women, also contributes to the lack of illustrations of overweight Caucasian female sexuality. Moreover, *American Idol* also contributes to a media representation of African American women that features sexuality as a dominant trait. This representation dates back to the Hottentot Venus, an African woman displayed in nineteenth century Europe as an example of exotic African sexuality (Magubane, 2001). The *American Idol* representation of African American women, while overall more positive than negative, also contributes to the history of the oversexualization of women of color. This quality is detrimental for the representation of African American women in media and may negate the positive characteristics depicted on the show, namely love of family, religion, and sincerity.

Desexualized R-and-B Divas

The converse of the big, beautiful, and overtly sexual diva was the desexualized diva. These women, while of average size, appeared uncomfortable with their bodies and sex appeal. There were fewer desexualized divas than sexualized. The desexualized divas included second-season contestant Kimberly Locke and sixth-season contestant Melinda Doolittle.

Like the other divas, these women had big voices and religious affiliations. They also shared an affinity for rhythm-and-blues songs. However, they dressed conservatively. They often wore business-casual type of clothes: slacks, long skirts, and long-sleeved, high-necked blouses. Both Kimberly and Melinda often appeared uncomfortable with their looks and verbally expressed it.

Season-six competitor Melinda was of average weight, not too heavy or slim, especially when compared to her fellow contestants on season six; Lakisha Jones and Jordin Sparks were both clearly bigger women. Despite this, Melinda seemingly lacked confidence in her appearance. Additionally, the judges generally praised Melinda's performances week after week. Each week, she was surprised when the judges complimented her.

Part of Melinda's discourse was self-deprecation. She was her own biggest critic. When working with guest mentor Jon Bon Jovi, she qualified her performance by saying "I am so bad at rock. I am just learning" (May 1, 2007). Prior to appearing on *American Idol* Melinda was a backup singer and she never truly appeared comfortable in the lead singer spotlight.

Despite having powerful voices and singing R-and-B songs, Melinda and Kimberly didn't have the confidence of divas. They did not own their archetype. This insecurity translated into a lack of sex appeal, which may have ultimately hurt their chances of winning *American Idol*. Both finished their seasons in third place.

Female Urban Singer

An offshoot of the female R-and-B diva was the female urban singer archetype. The urban singer is similar to the diva, and may have performed some rhythm-and-blues songs, but in general the archetype had more in common with hip-hop rather than R-and-B. These archetypal singers were youthful and came from blue-collar backgrounds. Season three contestants Camile Velasco and Leah Labelle fit this archetype. This urban archetype can be traced to female hip-hop progenitors such as Queen Latifah, MC Lyte, and singer Lauryn Hill. The female urban singer archetype was dissimilar from the R-and-B divas through song selection (fewer overt R-and-B songs), through visual presentation (no diva glamour dresses), and the absence of a focus on church or religion.

Camile identified as Filipina and hailed from Hawaii. She introduced herself on the show by saying, "Aloha, my name is Camile. I'm eighteen years old and I'm from Maui where I work as a waitress" (February 17, 2004). Her blue-collar background was firmly established through verbal discourse and visual images of her waiting tables at IHOP in Maui. Later, it was revealed that her parents actually owned the restaurant, diminishing her already-in-place blue-collar status.

At Camile's audition, she sang hip-hop group The Fugees' song "Ready or Not," emulating lead singer Lauryn Hill. The judges unanimously loved it. "Ready or Not" was a hit for the hip-hop group and singing it aligned Camile with an urban style of music. In the semifinal round, Camile strayed from this genre by singing a Brian McKnight song and the judges reacted poorly. Randy said, "What happened to my hip-hop girl, you that had the whole Lauryn Hill vibe? The tone was still there, but not the right song. You're like a young Lauryn; don't deny me that, man." Paula concurred, "All three of us said you are a young Lauryn Hill, that's what we were excited to hear." Randy asked her to "bring the hip-hop back, yo." Camile agreed and said, "I'll do it, baby" (February 17, 2004).

However, once in the final twelve, Camile neglected to adhere to her decided archetype. On the top-twelve performance night, Camile sang "Son of a Preacher Man," a Dusty Springfield song. Paula commented, "We compared you to Lauryn Hill. That vibe is not coming out tonight" (March 16, 2004). Eventually, Camile finished season three in ninth place. Her initial audition securely placed Camile in a previously unexplored archetype. The show discourse and air time given to Camile seemed to indicate that the show (as communicated primarily through the judges) seemed to approve of and wanted to see more of this archetype. Camile ignored her archetype, confusing it by choosing unrelated songs.

Male Urban Singer

The corollary to the female urban singer archetype was the male urban singer archetype, typified by season-six contestant Blake Lewis. Blake embraced hip-hop and beatboxed his way through several songs, thus creating his own *American Idol* archetype. He was Caucasian and attractive, and he attempted to give almost every song he performed an urban hip-hop twist through his beatboxing. Whereas Camile never fully committed to the urban archetype, Blake enthusiastically embraced the male version, building a more developed male urban singer archetype than the female version. Blake was the sixth-season runner-up to winner Jordin Sparks.

Before entering the *American Idol* competition, Blake was an aspiring musician in Seattle, Washington. At his audition, his beatboxing impressed Paula and Randy and informed those watching of his distinctiveness from previous contestants. The discourse about Blake established him as a fresh, innovative contestant, a new show archetype combining R-and-B elements with his hip-hop influenced technique. As Paula said, "You're making this competition hip and cool. You're the dark horse" (March 27, 2007). The following week she reiterated, saying, "You're a hip cat. You're just cool" (April 3, 2007). Later, again she said, "There ain't nothing old about Blake. Blake, you are the contemporary rebel in this" (May 8, 2007). Blake was the contemporary rebel because he dared to update *American Idol* with the flavor of hip-hop. Not only did Blake beatbox in performances, but also the discourse reminded the audience of his skill and archetype. For example, Ryan introduced the season finale by asking who would win, "the beatboxer from Seattle or the sweetheart from Arizona?" (May 23, 2007). Perhaps the next contestant to follow in Blake's footsteps will attempt to add rapping into song performances.

While rap and hip-hop are very popular, they were relatively absent on *American Idol* in comparison with other types of music. *American Idol* is a singing competition; although many rap and hip-hop artists are talented, they may not have the singing skill required for the program.

Blake's archetypal affiliation was clinched during the season-six finale on May 23, 2007. Various guest musicians performed with contestants. For example, Melinda Doolittle sang with gospel great Cece Winans. Blake performed with rap legend Doug E. Fresh, which was the first time rap music appeared on *American Idol*. Together, Blake and Doug beatboxed for a relatively quiet theatre audience in a display that previously would have been very unlikely on *American Idol*. Pairing Blake with Doug E. Fresh legitimized Blake's urban archetype, giving the archetype hip-hop's figurative nod of approval. In a sense, Blake became the Caucasian, sanitized, family and pop-music friendly version of rap. It is unknown if an African American man would succeed in the urban archetype.

Male R-and-B Singer

There were two types of the male R-and-B singer, sexualized and nonsexualized. The sexual R-and-B archetype consisted of a usually slender, generally handsome, and typically African American man. This archetype included contestants EJay Day, Corey Clark, Charles Grigsby, Anwar Robinson, Nikko Smith, Chris Richardson, and Brandon Rogers, among others. The nonsexualized R-and-B male archetype, also typically African American, included contestants Ruben Studdard, Rickey Smith, Scott Savol, George Huff, and Chikizie Eze. These men were overweight, could be perceived as unattractive, or perhaps not heterosexual, and were not discursively positioned as sexual.

Sexualized R-and-B Singer

The sexual R-and-B singer was repeatedly referenced as attractive or as a heartthrob. This archetypal character chose R-and-B songs to perform, or made pop songs sound as if they were in the rhythm-and-blues genre. Historical antecedents include Marvin Gaye, Sam Cooke, Smoky Robinson, and Teddy Pendergrass.

The show discourse reiterated the handsomeness of these men. According to Paula, when season-four contestant Anwar Robinson smiled, "it melt[ed] America's heart" (April 5, 2005). The visual images corresponded with the verbal discourse. Most of these men would be considered to be very good-looking in American culture and dressed stylishly. They did not look similar to each other, but all were positioned as attractive. For example, Anwar Robinson had long dreads, while other men had short hair. They wore sexy clothes, often with the top few shirt buttons undone showing their chests, or tight shirts and pants. The discourse emphasized their sex appeal through clothing, comments made by Ryan or the judges, and showing female fans screaming.

The male sexual-singer archetype contestants all sang R and B or soul songs. Artists they covered included Earth, Wind, and Fire; Dionne Warwick; R. Kelly; Chaka Khan; Marvin Gaye; Stevie Wonder; Ray Charles; and Sisqo. Additionally, the discourse of the show applauded their selections (even if not their performances) and encouraged them to continue singing similar selections. As Randy said to season-six contestant Chris Richardson, "Don't be afraid to jump to your R-and-B side" (March 27, 2007). Randy urged Chris to maintain and supplement his R-and-B archetype by continuing to sing R-and-B songs.

Nonsexual R-and-B Singer

The nonsexual R-and-B singer discourse omitted flirtatious behavior, sexual banter, or any mention of romantic relationships. In addition, while some of the men were not explicitly ugly, physically most of these men were not typically handsome. Their song choices were consistent with the sexual R-and-B archetype.

Ruben Studdard was the first contestant in the nonsexual R-and-B archetype. The velvet teddy bear metaphor Gladys Knight bestowed upon him constructed Ruben as cute, non-threatening, and affable. The show repeated the teddy bear metaphor often. Ruben himself did resemble a smiling teddy bear, soft and squeezable, and was friendly and approachable in his jeans and ever-present jersey. His amiable and casual demeanor, as well as oversized frame, made him nice instead of sexualized.

Other contestants, such as season-seven contestant Chikizie and season-three contestant George Huff, were not positioned as cute, nor were they sexualized at all. George may have been heterosexual or homosexual, but a viewer would never know—the show discourse situated George as asexual. George rarely flirted, acted in a sexy fashion, or mentioned the possibility of a significant other. This is a direct contrast to the overtly sexual heavy R-and-B divas.

The asexual men were often overweight and, analogous to the R-and-B divas, they frequently mentioned religion or church and were often family-oriented. The nonsexualized men wore bland clothes, such as Ruben's jeans and jerseys, or, when they did attempt to be sexy, merely seemed boring or average. George Huff, for example, often wore clothes that may have been construed as sexy if someone else was wearing them. George, however, looked as if he was trying to be attractive by wearing clothes that never really seemed to suit him.

By omitting any discussion of these men's sexuality, *American Idol* promoted heteronormativity. Instead of facilitating a national discussion about different types of sexuality, all contestants were presumed to be heterosexual.

This lack of sexual discourse also removed the asexual men as a hypothetical sexual threat to the audience. Statistically, African American men on television are most likely to be depicted as criminals, particularly in the news (Chiricos and Eschholz, 2002). Large African American men being portrayed as asexual mitigates the audience possibly associating them with frightening or potentially sexual criminals found on the news. *American Idol* broadens the representation of African American men on television through its in-depth focus on personalities. However, the asexual African American men are missing an essential component of their personas; this eliminates some of their humanity while also reinforcing heteronormativity.

Blue-eyed Soul Singer

An offshoot of the male R-and-B archetype was the blue-eyed soul singer, typified by season-five winner Taylor Hicks. Blue-eyed soul is a common phrase in the music industry that denotes African American soul music, such as Motown hits, sung by a Caucasian person (Garofalo, 1994). Well-known blue-eyed soul artists include The Righteous Brothers, Dusty Springfield, Joe Cocker, and duo Hall and Oates. Taylor, and to a lesser extent his fellow season-five contestant Elliot Yamin, fell squarely into this archetype. Taylor had musical talent, sang soul songs, and was identified through discourse as a white man singing soul, a new *American Idol* archetype. Taylor was never referred to as a rhythm-and-blues singer, but rather as a soul singer.

Taylor Hicks was white, of average build, and had shockingly gray hair in a short, non-descript haircut. He frequently referenced his fan base, called the soul patrol, shouting "soul patrol" during performances, voting, and basically whenever he could work it in. Over the course of the season he sang songs by Stevie Wonder, Sam Cooke, Otis Redding, Joe Cocker, and Elvis Presley. The discourse of the show repeatedly referenced him as a soul singer. For example, Taylor was "ecstatic" to sing Elvis' "In the Ghetto" because "I've always wanted to sing this number . . . it's one of the more soulful numbers that Elvis did" (May 10, 2006). Fittingly, during the broadcast of Taylor's visit home on the top-three competition evening, the song "I'm a Soul Man" (originally recorded by Sam and Dave, most well-known as a James Brown hit) played continuously under the spoken dialogue.

Taylor also had his own spastic dance, reminiscent of Joe Cocker's dancing and also James Brown's stage presence and persona. Taylor's stage presence was clearly strong; his unusual soul-music dancing seemed to captivate the audience, who responded by cheering and dancing themselves. James Brown's theatrical, over-the-top, high-speed feet movement is the clear ancestor of Taylor's stiff, shoulder shrugging and bunching, leg-tapping boogie. Taylor's dancing, and the resultant stage presence, was a unique element to Taylor's blue-eyed soul archetype.

Male Country Singer

An additional archetype throughout the first seven seasons of *American Idol* was the male country singer. Season-two contestant Josh Gracin originated this archetype on the show. Discourse concerning Josh also featured an emphasis on his military association, a trait shared by another country singer, season-six contestant Phil Stacey. Season-five contestant Bucky Covington inhabited the country archetype without any military association.

Josh's musical preference was country music. He sang songs by Garth Brooks, Lonestar, and Edwin McCain. However, he also sang other songs in a country fashion. On March 18, 2003, Josh sang "I Don't Want to Miss a Thing" by Aerosmith. Josh made a rock-band ballad sound country through musical arrangement, the twang in his voice, inflection at the end of words, and a seemingly southern accent.

Not only did his musical preference help label him as the country archetype, but his physical appearance and dress did as well. Josh was Caucasian, had a Marine-issued buzz cut, and appeared somewhat physically fit. His typical performance outfit was jeans, a button down shirt, and cowboy boots. In every show, Josh also wore his Marine dog tags around his neck, consistently reminding the audience of his active military status. Members of the military in the audience were also shown during Josh's performances. Furthermore, during judging Josh always stood stiffly, arms behind his back, in what seemed like a military manner.

The season-two discourse situated Josh as a country singer and as a Marine. At his initial audition, he identified himself as a lance corporal in the Marines. During the semifinal round, he elaborated, "I want people to see me as a down-to-earth kind of guy, and who I am, a husband, a father, a marine, and hopefully an overall great singer." Simon noted after Josh sang, "Josh, the marine. You're getting a lot of publicity" (February 25, 2003).

Later, on the night the top seven contestants performed the music of Billy Joel, Josh himself remarked on his country archetype. He said:

> To me, nowhere in the *American Idol* title does it say American pop idol, American R and B idol, American country idol. So, I wanted to show the fans that I can bring my country side with my pop side together and give them something that they've never heard from me before. (April 15, 2003)

Josh attempted to combine pop and country, formulating a new aspect of his personality, by singing the Billy Joel song "Piano Man." The judges generally liked it, although Simon found nothing memorable about it. The following week, Josh sang the Diane Warren-penned ballad "That's When I'll Stop Loving You," a song recorded by boyband N Sync. Simon negatively commented on Josh's performance, saying that "a few weeks ago you were going to go country and you've just changed your whole vocal style again" (April 22, 2003). Josh later stated that he thought: "I've sang country since week eleven, this is a chance to get out and do something different . . . I didn't choose it wisely" (March 22, 2007). Deviating from a contestant's constructed archetype by singing a different musical genre usually dilutes the strength of the archetype, as Josh himself noted. However, Josh was already established as a country singer and thus remained firmly within the country archetype. More country-tinged performances would have enhanced his archetype.

Other Male Country Singers

Season-five contestant Bucky Covington was also clearly within the country archetype. Bucky was from North Carolina and had a thick southern accent. He had flowing, shoulder-length blonde hair, wispy facial hair, and imperfect teeth, and he wore country-influenced outfits, usually jeans.

The show discourse repeatedly referred to Bucky as Southern and a country singer. As Randy said, "Dude, I like that you're doing the South proud" (February 22, 2006). The next week, Randy reiterated, "Bucky, Bucky, like me, man, representing the dirty South. You know what, this was a great song choice for you, kinda the whole Garth Brooks, Bob Seeger kinda thing, kinda rockin' country, kinda exactly who you are, man" (March 1, 2006).

The following season another military country singer emerged. Season-six contestant Phil Stacey had spent time in the navy. Another element that linked Phil and Josh was family. Phil spoke often of his daughters, as had Josh. Phil also sang country songs and discourse identified him as fitting the country archetype.

Phil didn't emerge as the season-six country archetype until the evening the top seven performed. That night, Phil sang "Where the Blacktop Ends" by Keith Urban. Phil explained his choice thusly: "The reason that I picked the song is because it reminds me of growing up in a place like Kansas and really just getting to and playing in the dirt and being a country boy" (April 17, 2007). The judges all responded positively to Phil singing country music that night. No other contestant that season had identified himself as country. Randy said in surprise, "Who would've thought, definitely not me. From an accomplished producer, you're going to have a career man in country music. That was really good. Who knew? I didn't know. Now I know Phil Stacy." Paula agreed, "It does seem like this is the genre you feel most comfortable in. . . . You sounded good from the beginning to the middle to the end." Lastly, Simon concurred: "This is the only time I heard you in a long, long time where I thought you chose a good song. I thought you sounded good and even more surprisingly, we actually saw some personality" (April 17, 2007). The following week Phil sang another country song, this time "The Change" by Garth Brooks. Simon continued his line of thought from the previous week:

> Last week we heard more of kind of a country tone in your voice, which I think suits your voice much, much more than you sang it tonight. I gotta tell you, if you can connect the two elements, you actually could do very well in this contest because I think people like you. (April 24, 2007)

Simon advocated that Phil continue to sing country music and actively seek the country archetype. Ryan summed up Simon's thoughts: "I think he's saying go for that country market man; you're going to be very successful" (April 24, 2007). Ryan's comment expressed that each contestant can fill a certain role, reaching a particular audience. A believable archetype appeals to a certain audience, thus possibly boosting popularity and lengthening a contestant's time on the show. That particular season, Phil was the only contestant attempting to fill the country archetype and therefore could appeal to all country fans. In fact, after finishing the season in sixth place, Phil went on to release a country music album.

Female Country Singer

The complement to the male country singer was the female country singer archetype. The first female country singer archetype also appeared in season two, in the guise of Carmen Rasmussen. Subsequent female country singers included Carrie Underwood, Kellie Pickler, and season-seven contestant Kristy Lee Cook. Carmen, Carrie, Kellie, and Kristy sang country music, physically appeared as country stars, and were discursively positioned as country singers. Earlier incarnations of the female country singer include Norma Jean, Kitty Wells, Patsy Cline, Dolly Parton, and Loretta Lynn.

Season-two contestant Carmen Rasmussen initiated the *American Idol* female country archetype. She was blonde, beautiful, and a youthful seventeen years old. She sang songs by LeAnn Rimes and Martina McBride, and made other songs sound country through arrangements and vocal delivery. An example of the discourse surrounding Carmen occurred on the March 18, 2003 episode, when Simon noted to Carmen that "you're a great country singer." Carmen also looked like a country singer, wearing jeans, cowboy boots, and other similar outfits.

While Carmen was the first *American Idol* female archetypal country contestant, season-four winner Carrie Underwood is the best known. Like Carmen, Carrie was blonde and beautiful, discursively positioned as a country performer, and sang country music. Carrie called herself a "farm girl," a "sweet and innocent girl from Oklahoma," (March 1, 2005) and a "country singer" (March 22, 2005). Paula added to the discourse by saying, "I love when you sing country, you just feel so natural singing those songs" (March 29, 2005). Later, on the top-three contestant performance night, Clive Davis complimented Carrie, saying, "Country pop is your element, it just radiates from you, not only in your voice, but your face and your body. You really nailed it, congratulations" (May 17, 2005). No matter who was speaking, the verbal discourse, as well as her visual appearance, marked Carrie as the quintessential *American Idol* female country singer.

Other Female Country Singers

Other female country singers included Kellie Pickler and Kristy Lee Cook. Both identified as being from rural areas and growing up in modest homes. Physically, both were blonde and attractive, just like Carmen and Carrie. Interestingly, Kellie's home life resembled a country song; she had the ultimate sob story of being abandoned by her mother and being raised by her grandparents while her father was imprisoned. Kristy Lee Cook's discourse focused more upon her rural background. Both Kellie and Kristy sang country music and were discursively placed within the country archetype.

At her initial audition Kristy Lee Cook sang "Amazing Grace." Simon commented, saying, "I like the country tone to your voice" (January 15, 2008). Later, she deviated a little in her musical choices and Simon reacted by saying:

> My only worry with you at the moment is—I don't know how to label you right now. I don't know exactly what kind of singer you really, really are. I have a feeling if you were to go kind of a country route you're gonna find it easier to make a statement, which I'm not sure you did with that song. But you know, you have real potential. (February 27, 2008)

After Simon's comments, Ryan asked Kristy, "What do you think about Simon's point when he talks about country being right for you?" She responded, "You know, I'm a total country girl, country singer, and if I'm here next week I'll prove it to him" (February 27, 2008). Kristy agreed with Simon's conclusion about her and resolved to demonstrate her dedication to her archetype.

And she did stick to her country archetype. She memorably sang country arrangements of The Beatles "Eight Days a Week," "God Bless the USA," and songs by Dolly Parton and Martina McBride. After her performance of Martina McBride's "Anyway," Simon noted that "Tonight, you look like a star, Kristy. It was smart. You're appealing to your audience. It was a smart choice of song" (April 8, 2008).

Singer-songwriter

A last *American Idol* archetype was the singer-songwriter. Of course, on *American Idol* contestants cover songs by other artists. However, in season seven (as mentioned earlier) contestants were able to accompany themselves on instruments. The resultant effect noted by the judges was that it seemed as though certain contestants were playing their own songs, even though the contestants did not write the songs. The two contestants most notable for the phenomenon were Brooke White and Jason Castro. Both presented the image of a singer-songwriter through repeated discourse and performances.

After one of Brooke's performances, Randy said,

> So, check it out, dog, Brooke. I like the fact that you brought the whole singer-songwriter thing. . . . At the very end I could feel you lose a little confidence. But you know what, I appreciate the whole singer-songwriter thing. (April 14, 2008)

The discourse throughout season seven also repeatedly linked Brooke White to famous singer-songwriters such as Carly Simon, Carole King, and

Tori Amos. The judges frequently mentioned these singers and Brooke as their peer, despite the fact that Brooke didn't sing her own songs on the show.

A similar discourse existed in reference to Jason Castro. Randy referred to Jason by saying he had a "singer-songwriter kind of vibe" (April 1, 2008). These types of comments recurred throughout the season.

Not only did the verbal discourse refer to Jason and Brooke as singer-songwriters, but visually they also appeared as singer-songwriters. Typically, Brooke would sit at a piano on the center of the stage, crooning a ballad as her fingers gracefully plucked out the notes. Likewise, Jason Castro appeared on stage in one memorable instance strumming "Somewhere Over the Rainbow" on a ukulele while singing the lyrics. Clearly, Jason did not write that song, which has been performed several times in different seasons on the program. He was covering a version of the song recorded by Hawaiian artist Bruddah Iz. However, visually he looked like a singer-songwriter, performing his own music without the full band or backup singers behind him. The simplicity of Jason, his voice, and the ukulele helped to initiate the singer-songwriter archetype.

CONCLUSION

The longer a contestant is on the program, the more screen time he or she has to develop his or her archetype. Presenting an identifiable archetype is essential to placing well in the competition. *American Idol* top-twelve contestants who did not present a coherent archetype, or adhere to one created in earlier seasons, did not last long in the competition. For example, season-four contestant Nadia Turner attempted to create a funk-oriented archetype. However, she came across as a mix of rock and rhythm and blues, never creating something recognizable as a single archetype. She finished in eighth place, despite being an early judges' favorite.

The *American Idol* discourse stressed "being who you are" and distinctiveness. This rhetoric contrasts with the very concept of archetypes, of fitting into a category and resembling others. In order to be successful, each contestant endeavors to appear both loyal to an archetype and individually different. Despite the prominence of the concept of individuality, the presence of archetypes is undeniable.

In general, the archetypal discourse omitted any mention of race, sexual orientation, or gender. In one perspective, this verbal omission erases these issues. For example, people of all colors were on the show, yet no one ever discussed race, what it means to be an African American, or to have to parents of different races. Thus, race is erased from the show itself through a lack of discourse concerning it, as is sexual orientation and gender. Even when Taylor Hicks sang soul music, a traditionally African American genre, race was never mentioned. Darling-Wolf (2011) found a lack of racial discussion with the same result in a similar program, the French *Star Academe.*

By omitting archetypal discussion about race, gender, sexuality, and class, only individual talent and work seem to factor in the rags-to-riches journey. Everyone is positioned as equal. On the show gender is also not discussed, and thus is also erased under the assumption that everyone begins at an equivalent level, although many archetypes were gendered in nature. Inequality of any type is never mentioned in the contestant discourses. If these inequalities, racism, sexism, xenophobia, and others, are not mentioned, then seemingly everyone has an equal chance of success.

However, while there was a lack of verbal discussion, visually the archetypes were typically composed of a single race. For example, there were no African American country singers, rockers, or girls/guys next door. While guy- and girl-next-door Justin Guarini and Jordin Sparks have African American ancestry, they were positioned as the Caucasian norm through an absence of discourse about heritage. The archetypal options for African Americans are apparently limited to the rhythm-and-blues archetypes. Perhaps the African American contestants themselves conformed to the established archetypes or perhaps the show edited their comments and worked to present them a certain way. The contestants and show most likely worked together choosing songs, honing performances, and shaping contestants' archetypal discourse. This may indicate that producers and others working on the show attempted to present contestants in a way they assumed the audience would find acceptable (or chose certain contestants who fit the archetypes). For example, all the contestants fitting the rock archetype were Caucasian.

There was not a Hispanic or Latino archetype visible. Overall, in the first seven seasons there were 82 top-twelve finalists (season one had ten finalists). Only three of those 82 mentioned any Latino heritage. One was eliminated fairly early in the competition: David Hernandez finished twelfth in season seven. In the first season, R.J. Helton finished in fifth place. While R.J. Helton did discuss his background at least once, the lack of discourse pertaining to his heritage, combined with the lack of people following in his footsteps did not help establish any sort of Latino archetype. Season-seven runner-up, David Archuleta, mentioned his Honduran heritage infrequently. Consequently, other aspects of his archetype were featured more strongly, placing him into the guy-next-door archetype, and reverting his race to the Caucasian norm. Similarly, the one Indian American, Sanjaya Malakar, also had other aspects of his personality featured much more frequently and in-depth, rather than his ethnic background. In contrast, the three Philippine

American women discussed their heritage, verbally discussing their Asian ancestry. This became a part of their archetypes.

Moreover, class was only marginally discussed through the context of a working-class background, which was a common denominator across all archetypes. Indeed, as having a humble background is necessary for the rags-to-riches narrative, it makes sense that all archetypes would include that element. In particular, the rock archetype emphasized the blue-collar background of the contestants. This was an effect the show worked to transmit; the audience never knew, for example, the circumstances in which David Cook grew up. The show presented him as a working-class bartender. However, his vocabulary and diction suggest an upper-class upbringing and education.

Many of the archetypes, specifically the girl/guy next door, country, and rhythm and blues, included a focus on small towns. The majority of the contestants were portrayed as being from small towns, including three top-five finishers who hailed from Birmingham, Alabama: Ruben Studdard, Taylor Hicks, and Bo Bice. Contestants often claimed to love small towns. This focus idealizes small-town American life and is similar to Gans's (1974/2004) news ideal of small-town pastoralism. Both the national news (Gans, 1974/2004) and *American Idol* favor small-town values over those of big urban centers. Small-town pastoralism includes the romanticism of country towns, an idealization of individualism, and support of moderatism (Gans, 1974/2004). The *American Idol* archetypes include all of these ideals.

The archetypes also indicated sexual standards on *American Idol* through who was sexual and who was not. Caucasian men, Caucasian women, and African American women were often portrayed as sexual. Some African American men were sexual, but a significant portion of the African American male rhythm-and-blues archetype was asexual. These men were often overweight or not conventionally attractive. Furthermore, no one was presented as homosexual. By including only heterosexual or asexual people, the show promotes heteronormativity, the belief that heterosexuality is normal and a benchmark. Homosexuality is then considered atypical. Additionally, by not allowing the unattractive men to be sexual, the show also promotes a beauty ideal related to sex. Only attractive, heterosexual people were considered sexual, as indicated by their flirting, sexually natured comments, and references to dating. The show maintains television norms of minimizing homosexuality and associating conventional attractivity with sexuality.

As discovered through the discourse analysis of the first seven seasons of the program, *American Idol* archetypes contain race, class, sexuality, and gender ideas. As previous research indicated, archetypes strongly feature one aspect of human nature and consistently repeat (Lule, 2001). Archetypes appear in all facets of media, including *American Idol*. Thirty million people watch *American Idol* on a weekly basis—these people are represented through the different archetypes, with which members of the audience may identify.

American Idol grants more visibility to groups underrepresented on television, such as the economically disadvantaged, African Americans, and Asians. However, fidelity to certain archetypal standards limits the complexities and comprehensiveness of portrayals. In turn, this stifles the possibility of exploring issues of oppression and inequality. The *American Idol* archetypes represent American cultural standards and the current status quos of class, beauty, sexuality, and race.

Chapter 5

Narratives of Authenticity

American Idol demonstrates that contestants deserve to be stars due to inherent talent and authenticity shown in performances and through backstage exposure. This is established by inviting the audience backstage to witness and contribute to the transformation of an average person into a star. Through exposure to the backstage, audiences learn about the contestants' authenticity, making them familiar to audiences, and highlighting their emotions. Authenticity is comprised of emotional authenticity, adherence to an archetype, and talent.

American Idol obfuscates Rojek's (2001) dichotomy of a public persona/personal self. Reality TV, and *American Idol* in particular, make some personal aspects of the self common knowledge—the private persona is now public and celebrated. The program humanizes contestants by presenting private doubts and fears along with contestants' backgrounds, family, and talent. While archetypal aspects may be common across certain contestants, emotional authenticity and talent are distinctive qualities for each contestant. *American Idol* amalgamates confirmed fame and the authentic self.

Authenticity is necessary for the contestants to appear to deserve to be a celebrity. The program positions the audience as fitting Andrejevic's (2004) notion of a savvy audience. The audience partakes in the celebrity transformation process by observing the contestant narratives, judging, and voting. Emotional authenticity, likability, talent, and song selection are some of the factors that create the authenticity of the competitors. Authenticity, combined with celebrity confirmation (discussed in the next chapter), results in fame for each contestant through the course of the *American Idol* season. Not only must a contestant walk the red carpet, confirming celebrity status, but also he or she must appear authentically talented and emotional to succeed.

This chapter first notes the importance of talent for authenticity, then examines how the backstage area, where emotional candor and hard work are depicted, is constructed. The subsequent section of the discourse analysis evidence concerns emotional authenticity. Following that portion, a third section explores how song selection aids in authenticity construction. The fourth and final section of this chapter relates several examples of portrayed authenticity. A conclusion sums up how *American Idol* creates seemingly authentic celebrities.

TALENT

According to the program, being unique and talented is a large part of success on the show. As Randy stated on the very first episode, "I think the main thing really for me is uniqueness. Having a different unique style, a different sounding voice, and also just having phenomenal talent" (June 11, 2002). A contestant can't be exactly like contestants or celebrities who came before her—she must be a little different, a new voice. Ironically, to be successful on the program, each contestant may adhere to a particular archetype, yet change it just a little to be different enough from other singers. The discourse of the show stresses that each contestant should be singularly talented and unique, yet fit into a familiar archetype so audiences can easily understand the contestant's image.

Innate Talent

Every season, contestants perform with a screen behind them, showing different unobtrusive scenes. In the first and second seasons, the scenes were conspicuously all nature-related, either a beautiful sunset or sunrise, or a depiction of oceans, mountains, fields, clouds, or a similar picture. The nature scenes are a parallel to vocal talent. Both are naturally beautiful and inherently exist. Authentic talent is thus innate and biological. The show repeats this each season. The message being expressed is that talent is inherent while image is separately constructed. Many of the people who audition for the show seem to believe in themselves, yet the audience is told they lack authentic talent. Celebrity confirmation can come later, but talent is necessary to get a foot in the door.

The discourse throughout the first seven seasons of the show emphasized talent. On the first season, Tamyra Gray and Justin Guarani were early front-runners. Simon said of Tamrya's rendition of "A House Is Not A Home" by

Luther Vandross, "Honestly, Tamyra, one of the best performances on TV, I've ever, ever seen in my life" (August 13, 2002). Likewise, Justin said of his singing ability: "This is a gift that I've been given, and I get to share that gift with whoever's listening and that makes me feel good" (September 2, 2002).

In the second season, Paula said to Trenyce: "You're either born with it or you're not. You are born with it" (March 11, 2003). She reiterated this the next week, saying that it is "as if you've been born to do what you are doing right now, on that stage" (March 18, 2003). At the season-two top-32 audition Simon said to Ruben, "We're looking for a star—you are a star" (February 11, 2003).

Also in season two, contestant Vanessa Olivarez noted, "I want people to remember me being a lot of fun and as not just an image, but also a talent" (February 18, 2003). She wisely separated the different spheres of the show, knowing that while they may be confused, each is important to demonstrate. Talent is an integral part of authenticity, which cannot be created without it. Talent is inborn, while authenticity is built.

Sometimes the judges will see talent in a performance that doesn't resonate with the audience or that perhaps isn't repeated again. Paula spoke about season-five tenth-place contestant Lisa Tucker:

> I look at you and you are this precious little gift. You light this fire inside my heart. And you are poised and you have this gift. And either you have it or you don't. And from start to finish you are a star. (February 21, 2006)

Simon agreed with Paula, saying:

> And that's what it's all about. It's finding occasionally somebody special. You have the potential with your career that we might look back in five, ten years' time and go this was the night, we'll remember Lisa Tucker. It was that good. (February 21, 2006)

The show presents the idea that a natural gift connotes that vocal skill, and thus fame. The comparative rarity of the gift warrants celebrity.

Guest judges are used to bolster claims of talent by occasionally praising contestants as well. For example, Barry Gibb said to Jordin Sparks: "You have a gift. When you go to your high notes, you can make the choice as to whether or not they're hard and intense or soft and gentle." And to the camera later, Gibb said about Jordin, "This is going to be, I think, one of our greatest female recording artists" (May 8, 2007). Similarly, Dolly Parton observed about season-seven contestant David Archuleta that "I really think he has the voice to become a great, great singer" (April 1, 2008).

Emphasizing the innate talent of each contestant demonstrates that he or she deserves to be a star. Talent is an intrinsic part of contestant authenticity, necessary to be a celebrity. The discourse stressed again and again over the first seven seasons that vocal ability was compulsory. Not only is it obligatory, it is inherent, and cannot be created through work or counterfeited. While work and practice can hone and improve a voice, natural aptitude must originally be present. Emotional candor and archetypal narrative coherence are likewise required to establish an authentic portrayal.

BACKSTAGE AUTHENTICITY IN *AMERICAN IDOL*

An essential element of *American Idol* is the seemingly transparent process of celebrity creation. The authenticity of the contestants is fostered through exposure to *backstage*. Backstage access supposedly allows an unfiltered view of the contestants' true selves. Backstage is traditionally where performers may truly be themselves and not perform their roles (Goffman, 1959). Here, I consider backstage to be anything shown to the audience that did not take place onstage during the actual broadcast of the show. Contestant interviews, rehearsals, meetings with mentors, and footage of the contestants with their families or home are considered to be backstage. When contestants are not onstage performing, they are backstage supposedly being themselves, able to ignore expected behaviors (Goffman, 1959). Several elements combine backstage to create this authenticity.

The program often presents contestants demonstrably emoting throughout their seasons. The "moments of authenticity" are ways that audience members relate to participants (Hill, 2005). Showing emotion to viewers is achieved through interviews, packages about life before the show, and behind-the-scenes views. Hard work, happiness, familial moments, tears, stress, laughter, and joy add to each contestant's authenticity. This makes the contestants more "real" or authentic, paradoxically adding to their celebrity by being normal. Being ordinary is a component of achieving celebrity through reality TV. Yet, at the same time, supposed insight into the authentic character of a contestant is matched with admiration of vocal talent and positioning as a celebrity. Contestants are portrayed as similar to audience members in emotion, background, and other ways, but the contestants are superior—they sing on television. The authentic emotions foster identification and diminish the distance between a viewer and a contestant, while the vocal skill and celebrity increase the distance between the viewer and the contestant. Authenticity simultaneously cultivates both a relationship and distance, helping create celebrity.

Backstage

The opposite of being on stage performing is the backstage area, where rehearsals take place and emotions may readily pour out. Audiences don't know if the backstage they see is real. Backstage is constructed through selection of footage and editing to frame the contestants as the show desires them to be seen. What audiences understand as backstage, the transparency of *American Idol*, is still a construction created by what the contestant chooses to present and what the producers determine the audience may see. Even though the backstage presented to the audience may still be considered front of stage, themes and patterns associated with backstage access are discernable.

Auditions

The entire audition process of *American Idol* lets the audience in on backstage happenings. Before *American Idol*, no other show privileged the audience by showing what happens before the majority of the show, the competition itself, occurs.

On the very first episode, Simon Cowell explained what makes *American Idol* different from other shows of its ilk:

> I'm here to do a job and I'm gonna do something which I think is gonna be a shock to the American public. We are going to tell people who cannot sing and who have no talent, that they have no talent. And that never makes you popular. We are going to show the audition process as it really is. Because shows in the past have not shown the brutality of auditions. Auditions are horrible places to go. And I'm warning you now; you are about to enter the audition from hell. (June 11, 2002)

At the typical first *American Idol* audition shown on television, a prospective contestant sings for the judges. He or she stands in front of a backdrop, facing the three (or sometimes four) judges, who sit while the person sings a capella. It is unclear exactly how many people, cameramen, crew, etc., are actually in the room. After the person finishes singing, he or she momentarily waits in trepidation for the judge's immediate critiques. Occasionally, the judges praise the performance. More often, they crack jokes at the auditioner's expense, put down his singing ability, criticize his appearance, and generally insult the person auditioning. Reactions shown on television to this critique vary from meek acceptance to tears, confusion, and outrage. The typical auditioner functions as a punch line for the judges, who often laugh at each other's jokes. These initial auditions also provide a baseline for judging talent throughout each season.

Allowing audiences to view the audition process in all its brutality was novel. Previously, auditions were an experience that people outside of the entertainment industry (the majority of people) were most likely unfamiliar with. While Simon may have been considered overly cruel to people, in general, auditions are not fun or nice experiences. Most people auditioning for an entertainment production are rejected. Audiences privy to this process are immediately made show insiders. They see people succeed and fail and are able to form their own judgments about auditioners. Placing the audience in the room with the judges, laughing at the unskilled, immediately creates complicity between the show and the audience. Moreover, the audience is aware that they ultimately will be the judges, the ones who choose the winner.

Vocal Coach

The program also suggests audience backstage presence through the vocal coach. Debra Byrd is the primary vocal coach, listed in the credits as Vocal Coach and Arranger, and is always referred to by her last name. Contestants were often shown working backstage with Byrd and celebrity mentors. In the first season, Byrd was visible in several of the rehearsal scenes. She was not mentioned by name, but was noticeable standing next to practicing contestants. By season three, contestants were speaking her name to the cameras and crediting her in song development.

For example, season-three contestant Camile Velasco narrated images of her working with Byrd and cried, "During vocal rehearsals yesterday, Byrd told me I was kinda holding back. She pinpointed to a *T* exactly what I was doing. Just overwhelmed me with emotion, I couldn't help myself." Byrd said to her, "Let the whole world see your fabulousness, and your gorgeousness, and your great talent." Camile told the audience, "I feel like I walked out of there different than when I first walked in" (February 17, 2004). The show displays the contestants working hard backstage, so that their success seems earned.

While Byrd is shown in almost every other episode, usually flashed upon in a rehearsal scene, an example from season seven indicated the extent of her involvement with the contestants' performances. In this scene, not only was contestant David Cook working with Byrd on his song, but he was also standing on stage, blocking his moves for that night's performance while Byrd was directing him from the ground in front of the stage. David held a microphone stand and tried out different stances, looking to the people on the ground for encouragement or critique (April 8, 2008). Not only is Byrd helping with the vocal arrangement, she's also helping to coach performance choreography.

The guest mentors also work backstage with the contestants, helping shape the performances. For example, season-two guest mentor/judge Diane Warren, a noted songwriter, worked with contestants at the piano, coaching them through songs she had written (April 22, 2003). Diane is a behind-the-scenes person—not an artist but a songwriter, unfamiliar to audiences. Having Diane Warren work with and judge the contestants especially indicates the backstage nature of the program.

Seeing contestants and guest mentors working together helps illuminate the process of performance creation, adding to the feeling of backstage access and the visibility of the contestants' authentic talent and hard work. Other examples of songwriters and performers working with contestants include David Foster, Andrea Bocelli, Barry Manilow, Jon Bon Jovi, Neil Diamond, Neil Sedaka, and Dolly Parton. These esteemed musicians contain their own authenticity as well. The role of the guest celebrity is often more than just an appearance or performance. It appears as if many actually spend time working with the contestants, helping to hone performances, choosing keys and verses, and generally adapting the songs for the contestants. Contestant Elliot Yamin worked with Andrea Bocelli and said, "When he first wanted to implement the changes that he wanted to make to it, I wasn't really getting it at first, so it took a few tries, a few takes" (April 25, 2006). While the amount of time they actually spend together is unclear, audiences may assume that the songwriters or performers do influence and somewhat aid the contestants. While interacting with the guests and the vocal coach, the contestants are established as authentic, by having talent and needing to work hard.

Soundboard and Studio Setting As Cues

On *American Idol,* contestants, celebrity guests, and judges often sit in front of audio soundboards or in recording studios. The soundboard symbolizes the recording industry, the possibilities for commercial success related to *American Idol*, and the effort and work of the contestants and those who produce the show. The presence of this equipment also signals to the audience that this is backstage, allowing viewers to feel knowledgeable about behind-the-scenes action. Here, the audience is privileged to learn thoughts and emotions not possible to present on stage. The judges also give more in-depth performance analysis by the soundboard.

Beginning in season three, the show repeatedly utilized the soundboard and studio setting to connote and reinforce the backstage setting and music industry connotations. For instance, when Diana DeGarmo's audition aired, her performance was interspersed with the judges commenting while sitting at a soundboard. The presence of the soundboard placed them backstage (as opposed to onstage at the judging table) and because of this, it seemed as if they were giving off-the-record opinions (January 24, 2004). While Randy's opinion as an expert is a given, sitting at a soundboard also reinforced his trade as a music producer.

Season three also featured contestants singing in sound booths, recording songs that eventually ended up on the season-three compilation album "Greatest Soul Classics." A special segment on the making of the album provided album publicity, helped build contestant celebrity status, and provided a backstage look at the recording. This was the first contestant compilation album to debut while the show was still airing, entering the market on April 27, 2004. This segment evidenced the contestants' authentic talent by showing them singing offstage. It also featured Gladys Knight speaking about soul music and then the contestants in the recording booth discussing what soul music means to them, as well as singing the album songs. Additionally, the contestants spoke about their emotions and thoughts on the recording process. The segment culminated with contestant Matt Rogers holding up the CD case and saying, "Last time I was looking at something like this—you got Clay, you got Ruben, Kimberly, like all the *American Idol* greats are on there. Now I'm one of them! It's awesome!" (April 6, 2004).

Likewise, sitting near a soundboard, season-five contestant Bucky Covington spoke about meeting Stevie Wonder: "Me bein' from a little town like Rockingham and everything, stuff like this does not happen every day you know. It's not like I'll be sitting in a bar somewhere and Stevie's gonna come walking in" (March 14, 2006). The interview is constructed to seem as if the audience is backstage with Bucky, engaging in an unadulterated experience with him.

In a more in-depth example of the backstage work that goes into readying a song for performance, season-seven contestant Michael Johns was shown working out how to sing a Beatles song with the engineers and the band in the studio. Michael said:

> I'm actually doing a song called 'A Day in the Life.' It's one of my favorite Beatles' song. I think it's Lennon and McCartney's masterpiece. It's hard to take a five- or six-minute song and compress it into a minute or forty or so. We're taking all the best parts and I think we've got a pretty good arrangement. (March 18, 2008)

A voice then spoke from offscreen to an exasperated Michael, "You've already taken half of it [the song] already" (March 18, 2008). While speaking the narration, Michael was shown singing in a recording booth and working with band members to make a suitable arrangement for the show. This scene made visible to audience members the process each contestant goes through each week to prepare his or her song. This visibility makes each audience member an insider, helps create a strong connection between the audience and the show, reveals emotions, and is a unique part of the *American Idol* competition. Through the emotions, hard work, and talent illustrated, each contestant establishes that he or she is an authentic star, deserving of celebrity.

By season seven, the contestants were shown not just working on their songs to prepare for performances, but also recording their songs for release on iTunes. A package that highlighted this process was set entirely in the recording studio and aired March 26, 2008. Contestant Carly Smithson introduced the package by saying, "We are at the studio and we are recording our tunes for iTunes." While she spoke, images on screen included her singing into a microphone in a recording booth and reading lyrics. Jason Castro explained, "When we perform we have to cut down the song to about a minute and a half. Here at the studio we get to do a full-length version, about three minutes or so." Brooke White spoke in front of the window into a soundboard room containing technical equipment and technicians: "It gives the listeners a chance to hear a little more than they would get at the show." The next images shown were Chikizie listening to headphones and Ramiele Malubay sitting at a soundboard and then on a chair with her laptop.

Ramiele informed the audience that "I've downloaded David Cook, Michael Johns, and Jason Castro just 'cause I've liked some songs that they've done." Also discussing the use of iTunes, David Cook said:

> When I'm trying to figure out a song or figure out what I'm going to do with it, I usually will use iTunes as kind of a reference source. I'll go through the library that they have on the iTunes store and just kinda nose around and see what I can come up with. (March 26, 2008)

Images shown during his voice-over included him rehearsing on stage, in the recording studio, and at a soundboard. Brooke White talked about her emotions during recording: "I still get nervous to record, I really do. My heart is beating really hard." A technician spoke to her: "You sound great as usual, beautiful. Watch your pitch." Michael Johns said, "Basically, they want you to get as much live takes as you can," over footage of him singing, erring, and restarting. The live takes are evidence of his singing skill and industrious hard work.

Carly Smithson spoke about the technical issues: "There's like these lingo words. They have all different slang for stuff." A technician spoke to Carly: "I tracked you before, so if you want to go all the way through and then go back and punch in." Carly said into the camera, "And then you see this huge big desk and there's all these buttons and I doubt any of them even know what it does." Next, there were more shots of a soundboard, of technicians, and of David Archuleta singing into a microphone. David Archuleta continued the technical theme:

> What I like to do when I have headphones on, I like to cover one ear so I can hear what's being recorded and then I like to keep one off, so I can hear myself and how well I'm sounding normally. (March 26, 2008)

Kristy Lee Cook explained, "If I'm learning a new song and I don't know it, I download it onto my iPod and I listen to it until I memorize it." Syesha Mercado and Brooke were shown sitting in armchairs, listening to music on iPod headphones while reading lyrics.

In that segment, backstage is symbolized though the setting of the studio and the technical music equipment. The contestants explained their emotions, how they work, and a little bit of how they prepare some songs for the show. What appears as an innocent segment about them recording for iTunes functions also as a commercial for the songs, as well as iTunes and iPods. The backstage nature of this segment adds to the candor, personalities, and insight into the contestants. Additionally, being backstage continues the well-informed feeling of being part of the show, as well as enhances the authenticity of the contestants. Their seemingly wide-eyed wonderment adds to their emotional authenticity as normal people, yet their presence in the studio positions them as celebrities.

Hard Work

All of these portrayals of backstage emphasize the difficult and time-consuming work that the contestants perform. Work is a consistent theme in the discourse of the show. The contestants may have natural talent, but need to work to earn success. Hard work is a component of authenticity. Ryan Seacrest interviews the contestants about their offstage lives; these questions and answers provide a window backstage for the audience to understand the contestants' work when they aren't on television. For example, Ryan asked season-six contestant Melissa McGhee, "What's it been like the past few days in the top 12?"

She responded, "It's been insane; we've worked like 17-hour days and got no sleep. It's been unbelievable, great."

Ryan inquired, "Do you feel like a superstar?"

Melissa replied, "Not like a superstar, but it's been a rollercoaster that I want to get off of."

Ryan continued, "How's your voice?"

Melissa answered, "I've been gargling and swallowing some stuff, like olive oil, but it helps the dryness" (February 21, 2006).

In season six, Ryan asked Jordin Sparks, "How did you about feel preparing for two songs last night?"

She answered, "It was kinda just, like I better pick [th]em fast 'cause I only have a couple days to learn them. I really liked the song so it was kinda easy for me to memorize it. But it was kinda double the trouble" (May 9, 2007).

Similarly, Ryan asked Syesha to "explain what it takes behind the scenes to be here preparing for a show like this."

Syesha noted, "It takes a lot of hard work, preparation, sometimes you go on vocal rest, sometimes you go to sleep early. On top of that it's a really busy schedule" (February 20, 2008). David Archuleta also spoke about the labor of rehearsing, saying "It's been pretty crazy this week; it's been a blast. I usually practice early in the morning or late at night. I hope I didn't wake the neighbors. I get pretty loud sometimes. But I have to practice sometimes" (April 8, 2008). On the season-seven finale Ryan asked David Archuleta and David Cook, "How did you prepare for this finale tonight?"

David Archuleta replied, "Just rehearsed as much as I can, in all my spare time."

David Cook answered the question with a reply about spiritual labor, "Well, for me, it just started out, I have a tall order trying to keep up with this kid. Just trying to find the right energy for each of these songs and hopefully you guys will oblige" (May 20, 2008).

Labor is often stressed in explanations of song performance. David Cook explained,

> You really have to be on your game 'cause everybody knows these songs, so you really have to do it justice. A *B*-plus effort isn't going to cut it anymore. Eleven other people [are] gunning for the same spot as me and they're all fantastic. So, I got my work cut out for me. My goal is to leave it on the stage. (March 11, 2008)

The contestants' work ethics add to their authenticity as celebrities; they merit celebrity through their work, skill, and authentic personalities.

Fun Facts Revealed Backstage

Fun facts revealed through backstage access also add to authenticity. These fun facts are similar to what Gamson (1994) referred to as tidbits. Gamson detailed the tension between journalists and publicists over revelations of a celebrity's personal life. Journalists desire personal tidbits to enliven their stories, while publicists want to control the way celebrities are presented in the stories about them. These bits of personal information are valuable;

therefore publicists wish to control "scarcity to maintain its [tidbits] value in extracting exposure" (Gamson, 1994, p. 94). In other words, interesting information can provide publicity.

The *American Idol* fun facts work similarly. The fun facts humanize the contestants and offer a supposed glimpse backstage. The contestants are able to show off their goofy sides or another aspect of their personalities, fostering authenticity through being their supposed selves. For example, Jasmine Trias showed off her hula dancing, taught it to other contestants, and apparently had fun doing it.

In another example, season-six contestants were asked backstage what music each was listening to as they dressed and had their hair done in front of makeup mirrors. Melinda said gospel, while Sanjaya said some country and some blues. Blake chose hard metal band Incubus. Jordin professed to really liking guest performer Fergie's new single, a seemingly fortuitous coincidence. Other fun facts included Jordin answering a viewer question on May 15, 2007, asking her favorite song of all time. Her answer: teenybopper sensation "MmmBop" by Hansen. A viewer asked Blake a more random question: "If someone were to make a movie of Blake's life, what would be the title and who would play him?" His answer: "Jim Carrey would play Blake and the movie would be titled 'Organized Chaos'" (May 15, 2007). David Cook answered a viewer question as well: "If you could succeed at any other talent what would it be?" David answered, "I don't know if it's a talent, but I wish I was more organized" (April 2, 2008). Access to the contestants' thoughts facilitates a sense of knowing the contestants well. The contestants are also able to demonstrate realness through the fun backstage facts.

In an interview during the seventh season, Jason Castro revealed an interesting fact about himself—he loathes interviews. Jason said,

> One thing I think maybe some people would be surprised to know is that I hate doing interviews. (He is shown starting and restarting again and again during this segment). Interviews can be really awkward. I make faces and am weird. I'm not good at words. I just really like to sing and I like music and anything to do with music. (February 26, 2008)

Jason's fun fact made him seem more authentic through his dislike and errors.

Fun facts provide knowledge about the contestants' lives, enhancing the audience's backstage familiarity with the contestants, their emotions, and the show in general. The feeling of being authentically backstage is built through the audition process, the vocal coach's visibility, soundboards and studios as cues, the ideal of hard work, and revelations of fun facts. The audiences' backstage access creates authenticity for the show and the contestants.

EMOTIONAL DISPLAYS

Emotions, when earnestly exhibited, convey an authentic personality (Marshall, 1997). On *American Idol*, contestants demonstrate Hill's "moments of authenticity" (2005). Tears, laughter, anxiety, joy, confidence, humor, and other strong emotions impart sincerity and authenticity. Personalities differentiate contestants from each other, particularly as the presence of talent is universal in the finalists.

Emotional displays must appear authentic and be unmistakably clear on stage and off in order to succeed on *American Idol* and foster contestants' celebrity. Emotions are expressed through stage performances as well as offstage. Judges look for emotions and frequently comment on their presence, or lack thereof. As Paula explained to season-two wild-card contestant Nasheka Siddall, "You have to emote, you do have to make us feel engaged." Randy continued by saying, "I was waiting and hoping for a little growth in the emotion area [displaying more perceptible emotions] and connecting with us." Nasheka rarely exhibited noticeable emotions; this may have helped prematurely end her *American Idol* journey, despite her talent (March 4, 2003). Three important thematic personality attributes displayed again and again are confidence, nervousness, and humor. Anger, regret, and similar emotions are rare on *American Idol* after the initial auditions. Sadness is occasionally depicted, but less frequently than more positive emotions.

Performance Emotions

Emotions are required for performance authenticity as much as talent. A performance that lacks emotional authenticity is labeled lackluster, unexciting, and mundane. As Martina McBride, a season-six mentor, advised the contestants: "You have to connect with the lyric and just forget about anything and just connect with the story" (April 17, 2007). Connecting with the story, analyzing and understanding the lyrics, and then conveying emotions are imperative to establish authenticity. A performance appearing emotionally authentic can be truly moving for the audience and the performer, making the sentiment of the song clear to everyone.

The judges are quick to point out when performances are missing emotional authenticity. In season two, on April 22, 2003, Simon said to contestant Trenyce, "It's as if you wear a mask on stage. If you could be a bit more expressive you could do really well." Similarly, guest judge Clive Davis appraised season-four contestant Vonzell Solomon's performance: "You picked a true soul classic. You're a very, very good voice but you gave it a very pop, spirited, likable rendition. It was winning because you're talented.

To me, I think you missed the soulful essence of the song" (May 17, 2005). It seemed like a compliment at first, but Clive truly expressed that while Vonzell has a good voice, her delivery of the song lacked emotion. She "missed the soulful essence of the song" that could only be conveyed through imbuing a voice with rich emotions. Her performance lacked soul. Similarly, on season seven Kristy Lee Cook sang the Beatles' song "You've Got to Hide Your Love Away" written by John Lennon. Randy claimed, "The whole song should have felt more emotional." Likewise, Simon noted, "The problem is you're not a good performer. It's like musical wallpaper in that you notice it, but it's forgettable" (March 18, 2008). Without visible emotions that seem real, the performance is uninteresting and lifeless.

In season three, emotions were a noticeable factor in the performance competition. Jasmine Trias noted this in her journey-home highlight reel after not becoming a finalist: "I've learned about my stage presence and also learned about expressing your emotions when you sing. It's all about passion" (May 19, 2004). Through the discourse analysis of the first seven seasons, it was clear that numerous contestants had tremendous vocal ability, and thus the ability to transmit emotions was paramount. This was especially true in the third season, when Fantasia raised the emotional stakes during her performance of "Summertime."

Fantasia sang "Summertime" twice on season three, first on the top-eight-contestants performance episode, and then again on the season finale. On the movie-soundtracks themed episode, Fantasia sat on the stage and expressively sang the theme from *Porgy and Bess*. Fantasia's performance of "Summertime" was remarkable not just for her demonstration of vocal skill but also for the visible emotions. While she sang the sad lyrics, tears that appeared real rolled down Fantasia's face, demonstrating the veracity of the emotion she felt. Fantasia wore a simple dress and an understated hairstyle and make up, directing attention to the music, her voice, and her lyrical sensitivity. Her rendition was poignant and tender. After performing, during the judges' evaluations, she seemingly couldn't stop the tears streaming from her eyes.

Randy commented, "That is the single best idol performance on any season I've ever heard."

Simon observed, "Fantasia, you prove the point, with that performance tonight why we are critical of people who just sing well, because there was something magical about what you just did."

Fantasia responded: "I felt my song" (April 14, 2004).

The emphasis on expressed emotions during performances continued throughout the next seasons. On season five, contestant Blake Lewis sang the Beatles' song "Imagine" and Paula noted, "It was the first real sensitive emotional performance from you, Blake, that I've seen and that goes a long way.

I really enjoyed it." Simon agreed, "It felt sincere; the only problem was it didn't really go anywhere. It was little bit like that throughout (flat horizontal hand gesture). But, I think the most important part probably was the fact that you sang it with sincerity. So, good" (April 24, 2007). The judges' acclamation of Blake's sincere emotions indicated to the audience the importance of sentiments as a judging criterion.

The winner of season five, Jordin Sparks, also realized the importance of emotions in performance. Reflecting upon singing a song written by Barry Gibb, Jordin said, "I went in today thinking I have to show him [Gibb] the feeling that I feel in the song" (May 8, 2007). In a different episode, Jordin and Ryan discussed her expressive performance of the song "I Who Have Nothing." Ryan noted, "You have the look down, like you really do love the person on the other end. You want to grab them and pull him through the screen" (May 15, 2007).

In season seven, Brooke White elicited similar comments from the judges regarding her sincere and emotive performances. On the March 11, 2008, top-twelve-contestants performance episode, Brooke sang the Beatles' song "Let It Be," accompanying herself on the piano. Randy called it a "very heartfelt performance" while Paula complimented her for "picking songs where we can feel your heart, America can feel your heart. It's having that connection, it's this emotional connection that makes people fall in love with you." And even Simon agreed, noting:

> It was, again, one of the best performances of the night. I thought it was a brilliant choice of song. I think you've done it three weeks now. And actually it's believable. There is a difference between karaoke and actually making it believable, showcasing your talent, which you've done tonight. Great. (March 11, 2008)

While the judges spoke, Brooke wiped away tears.

Similarly, David Cook explained his thoughts about performing early in the competition, during the semifinals round, on February 26, 2008. He explained that a good performance has many qualities: "It's charisma, and it's performance, it's all these extra things that come along with being a good singer." That day, he performed an exceptionally emotional version of the Lionel Richie song "Hello." Each phrase was imbued with emotion linked to the lyrics of the song. Passion, sadness, and longing were clearly manifested in each word, through the quaver in his voice, the expression on his face, and even his body language. While the emotions were expressed very clearly, David constructed the emotions in an effort to appear emotionally authentic.

Still, the discourse of the show positioned David as a very emotional singer and yet also fitting into the rocker archetype. After the "Hello" performance, Randy called him the "emo rocker." Emo is music industry shorthand for emotional and is a type of rock music popular with teenagers. David was the emotional rocker. On the season finale, David sang "The World I Know," a little known song by rock band Collective Soul. After another performance that ended with tears in David's eyes, Paula stood and applauded, saying, "I looked at you up on stage and you're standing in your truth. And you're delivering unbelievable songs with integrity and originality and I truly applaud you, David Cook, I do" (May 20, 2008).

Nonperformance Emotions

While emotional performances are crucial to establishing an authentic contestant portrayal, emotions are also exhibited offstage in order to create a complete persona. The contestants reveal emotions in interviews, behind-the-scenes clips, and packages about home lives. Disclosing fear, joy, confidence, or other emotions adds to the contestants' authenticity. Season-three contestant Latoya London pointed out that, for her, truthful emotions are difficult to falsify: "I'm very reserved. You know when my personality comes out. I can't fake it, it just happens" (May 3, 2004).

On Jordin Spark's season-six visit home, tears were in her eyes as she cried, "Who am I to have this amazing experience? I couldn't ask for anything more" (May 16, 2007). Fans cheered and screamed as she toured her hometown. Viewers at home witnessed her emotional turmoil at reconciling her sense of self with the reality of fame and fan support. Audience members could empathize with her confusion and humility. She seemed emotionally truthful.

Season-seven runner-up David Archuleta was an extraordinary case study of emotional revelation. Each week, female fans screamed in the audience and the judges were principally complimentary. David enhanced his appeal by refusing to acknowledge his talent and consistently showing surprise at his popularity. On February 19, 2008, in a semifinal round, Simon judged that David's performance "was, by a comfortable mile, the best performance of the night so far." David reacted by humbly shaking his head no, covering his eyes, and having great difficulty verbally expressing himself. He appeared shocked and said, "Wow, I wasn't expecting that. Wow."

Jumping onstage, Ryan said, "You're so likable. You seem a little shy after the song but during the song, very confident and comfortable on stage."

David replied, "I love it, especially here with the audience." Girls and women in the audience screamed and cheered. David said, "Oh man," and seemed surprised and happy by the comments and audience approval. David was able to balance confidence and ostensible humility. His performances resounded with confidence while his manner was meek and unassuming, joyful for received praise and applause. He professed to enjoy singing, yet seemed surprised by his natural talent and success, and therefore a little naïve. He ducked his head, smiled happily, and appeared too stunned to form sentences. A typical performance resulted in flattering approval from the judges and David's typical reaction. Ryan said to David, "You look stunned." David replied, "I couldn't believe that." And from the audience, multiple voices screamed in unison, "We love you David!" (March 18, 2008).

Confidence

Of all the emotional attributes expressed on *American Idol*, confidence was often communicated and stressed as important to have within the show discourse. Confidence is a converse to humility. Self-assurance is expressed in the way contestants perform, answer questions, and in all aspects of their emotions.

In an example, Simon said to Latoya London: "You are, I think, the best singer in the competition. I really do believe that. Let me ask you a question. Do you think you can win this competition?"

Latoya answered, "I definitely think I can win."

Simon rejoined, "Yeah, cause I think you now believe you can. And it's beginning to show. You did great."

Ryan agreed, "That is the attitude you have to have at this stage. Wow" (April 20, 2004).

Part of being authentic, being your supposed true self, is believing in yourself and being self-assured enough to share yourself with the audience. The show emphasized that confidence is necessary to succeed on the show. Confidence is a requirement for celebrity. Confidence usually begets poise and aplomb, allowing dignity in contrived situations and self-composure under duress.

In another example, Simon complimented Elliott Yamin after a performance, "I think you've grown in confidence. I wondered last week whether you actually felt that you deserved to be here. I think I see a different you walk on the stage tonight. You started to believe in yourself" (February 22, 2006).

Contestants themselves recognize the need for confidence. Brooke White noted that: "If I'm being completely honest about my biggest obstacle, it's confidence" (February 12, 2008). Understandably, the amount of confidence a contestant possesses can go up and down over the course of the season, contingent on comfort with the songs, physical well-being, and judges' commentary. Ryan asked Jordin a viewer question, "What have you learned about yourself during this whole idol process?"

She replied,

> I guess it would be that I can handle a lot more than I thought I could, I mean there's so much being thrown at us, two songs, doing those little music videos that we do, plus three hours of school on top of that. So I can handle a lot more than I thought I could. (May 8, 2007)

Nervousness

Nervousness is another emotion commonly displayed on the show. The show expresses nervousness to highlight the realness of the contestants and the reality of the competition. Season-three contestant Amy Adams explained:

> You're on it but it feels like it's not real. There's gotta be a part of you that feels a little bit scared and a little bit nervous no matter what. Thirty million people are watching this show and that's always in the back of your mind but you can't let that affect you. You have to let, you know, you be you and be ok with just being you and if America loves it, fabulous. (February 24, 2004)

Being nervous is expressed again and again on the show. The stakes of the competition are very high, the program is live, and there are millions of people watching at home and critiquing the performances. In addition, the immediate judging may be intimidating.

Many contestants expressed their anxiety. Taylor Hicks said, "I'm nervous as I could be," but once informed that he made the top twenty-four, he was clearly elated, as he demonstrated with a big smile, harmonica playing, and spastic dancing (February 15, 2006). Later in the competition, Taylor reiterated his anxiety: "I am nervous. I don't know what to expect" (May 17, 2006). Taylor's top competitor, Katherine McPhee, felt similarly, responding to the question of how she prepared differently for this show with, "I don't know. I was really, really nervous and my dress rehearsal wasn't as the way I wanted to go. I don't know, I don't know" (March 14, 2006). Katherine could barely utter a coherent thought.

Lots of little things can go wrong as well. Ryan said to Lakisha: "You looked a little uneasy last night. Why were you so nervous?"

Lakisha responded, "I had a lot of things that happened earlier that day, and I didn't have on what I was supposed to have on, and the key change, and over analyzing everything. I was hoping that Simon would kiss me again" (May 9, 2007).

Contestants also seem to worry about what has gone wrong and what can go wrong on the live show. As David Archuleta pointed out, "It's a huge opportunity to be on the show, and you don't want to mess it up! Hopefully, I do the right things. I'm getting shaky just thinking about it" (February 19, 2008). He continued the theme a few weeks later, saying, "This is definitely the most nervous I've ever been, just because it's a little scary doing this song, it's a difficult song for me. Uhhhh, sorry, I'm just a little stressed thinking about it" (March 11, 2008).

The nervous emotions featured repeatedly throughout the discourse humanize the contestants. While celebrity elevates contestants to a social standing above the audience, anxiety brings them down to audience level. Common emotions make the contestants seem sincere and real.

Humor

Humor augments emotional authenticity and also creates distance, enhancing the celebrity status of the contestants. When contestants exhibit humor, they create another seemingly authentic facet of their personalities, showing quick intelligence to be able to make jokes under pressure and allowing the audience to laugh. Humor may lengthen a contestant's stay on the program by adding depth to his or her character. At the same time, being funny is not a trait everyone has—it is an innate skill often possessed by celebrities. Performing on television, getting laughs, and bonding with judges and other contestants place the contestants farther away from audience members. Nevertheless, an important accomplishment achieved through humor is the revelation of a little bit more of a contestant's perceived authentic personality.

Season-three contestant Amy Adams was known for her humor. Randy asked her, "Are you a little nervous?"

Her answer: "A little. Okay, a lot." That was an amusing answer to a question almost everyone receives and seriously answers. As Paula pointed out in the same program, "You're a goofball with a nice pretty voice. You're just a goofball."

Amy replied, "Thanks, I think" (March 16, 2004).

In the same season Latoya London described herself as a young girl, saying, "I was a character, I was funny. I had a lot of impressions that I would do." She then performed a series of comical impressions, including the character Fire Marshall Bill from the television show *In Living Color*, the character Sheneneh Jenkins from the television show *Martin*, and a monkey (March 23, 2004). She showed the audience that she could be funny, that there was more to her than simply vocal skill and performance.

Season-four contestant Mikalah Gordon had some choice jokes that may not have been very appropriate for a family show. After receiving her golden ticket to Hollywood, Ryan asked her, "But what does this good fortune mean?"

The teenage Mikalah replied, "It means that now maybe I can buy my mom those implants that she always wanted" (January 26, 2005).

Wholesome Carrie Underwood won season four, while Mikalah was voted off eleventh. However, Mikalah's humor was not totally unappreciated, as she often hosts ancillary *American Idol* shows and made an appearance on the season-seven finale, hosting the hometown party for David Cook fans.

Unlike Mikalah, season-six contestant Chris Sligh was very funny in a wholesome way. At his audition, the overweight, curly-haired Chris said,

> Some people say I look like Jack Osborne, some people say I look like Jack Black. But when I look in the mirror every morning, it's not those people that I see. It's Christina Aguilera that I see. I think that's who my body looks most like. (January 30, 2007)

And after receiving his golden ticket, Chris noted that, "I let my personality show. I looked right at Paula and she crumbled in my chubby little hands" (January 30, 2007). For the contestants, innocuous self-deprecating humor is permissible, while blue humor is not. Even though the contestants are often sexualized, they are still a large element of the family-friendly show. Sexualization is allowed; jokes mocking it are not. While contestant sexuality is emphasized through dress and flirtatious behavior, contestant humor mocking sexuality was rare. The show has funny moments and is designed to entertain; however, the sexuality of the contestants is apparently serious business. Indeed, sexualization is a primary component of most pop stars.

While humor augments emotional authenticity, it also helps generate another essential element for contestants: likability. The show attempts to build likability for the constants. Contestants want to be liked, and some realize that a sense of humor may help mitigate poor reviews. After season-six contestant Phil Stacy received negative comments on a performance, he told the audience: "My daughter just got a stuffed animal cow and she named it Simon Cowell" (April 10, 2007). One-liners may provoke laughter or groans, but more importantly, they stimulate likability. Audiences like people who can make them laugh. Through emotions, the contestants seem authentic, average, and likable.

Likability and Ordinariness

On *American Idol*, there is a constant tension between celebrity and ordinariness. The program and the contestants endeavor to create a portrayal of the contestants as normal people. Audiences can relate to and like normal contestants.

Likability is a combination of friendliness and being average. The judges frequently highlight likability in discourse. Simon noted that season-four contestant Vonzell had the "likability factor" (May 17, 2005). Similarly, the next season Simon said to contestant Ace Young at his audition, "I don't think you're near the best singer we've seen. I like you. I think you're very likable. It's more about your personality than your voice" (January 18, 2006). Later in the season, Ace asked the audience, "What's up, everybody?" The audience cheered in response (March 14, 2006). Ace acknowledged the audience, engaging with the people and simultaneously showing a personable nature, stimulating likability and a connection with the audience.

A contestant often emphasizes an aspect of his or her character that is applicable to a lot of people, thus categorizing him or her with thousands of other people. Season-five contestant Chris Daughtry explained he shaved his head because he was going bald, a common ailment people can understand. Along the same lines, season-five contestant Kellie Pickler explained her love for dogs, just as season-four contestant Carrie Underwood stressed her love for all animals. Season-five contestant Melissa McGhee, a self-described "huge car freak," professed her love for cars, "especially fast ones" (March 7, 2006). Season-seven contestant Kristy Lee Cook, upon being complimented on her toenail polish by Ryan, happily replied, "Thanks, I did it myself" (April 1, 2008). Equally as broadly relatable, season-four contestant Bo Bice mocked his cool image by noting that if he had to sum up his "dancing in three words, it would be horrendous, terrible, and dreadful" (April 19, 2005). Then he purposely danced poorly, provoking laughter. Bo was funny and relevant to all people who doubt their dancing skills. He made himself more likable by allowing people to identify with him.

The contestant who epitomized likability was season-five contestant Kellie Pickler. She was a self-proclaimed "fish out of water" in an entirely new situation. She tried new things, mispronounced words, and claimed to have never had a vocal lesson, let alone sung in public. Kellie was a small-town girl happily lost in Hollywood. Her naivety, congenial nature, and willingness to make mistakes was very appealing and at times, amusing. She explained:

> I feel like a fish out of water here, but you know I've had some good times and tried some new things, had my first spinach salad, it was okay, kinda tasted like picking a leaf off a bush. But, um, I tried my first squid, they call that calamari, and that was—I could live without ever eating that again. (February 21, 2006)

Her lack of pretention seemed winsome. She wasn't attempting to seem sophisticated or knowledgeable; she even admitted that she didn't like the fancy food. Simon explained is thusly:

> Kellie, I don't know what it is, but you just cannot help but like you. If I hadn't heard the calamari (said in fake southern accent) bit and if I didn't know you and I just heard the song and I wasn't watching you, I'd go, it wasn't the best vocal I've heard. It was very rough round the edges in parts, but—the calamari is worth a lot of votes. Because you are cute, you're likable. I think America, if they haven't already, are going to fall in love with you. (February 21, 2006)

A few weeks later, on March 7, 2006, Kellie and Simon had an exchange that demonstrated her effect on people and her straightforward, simple nature. Simon tried to flirt and Kellie's lack of vocabulary stymied him.

> Simon: "Kellie, you are what's known as a naughty little minx."
> Kellie: "A what?"
> Simon: "A naughty little minx."
> Kellie: "What's a minx?"
> Simon: "A minx, it's kinda like a nice, bad girl. I really, really like you. I don't think you're the best singer, but you're interesting. I kinda prefer you to last year's winner actually [Carrie Underwood]."
> Kellie: "I'm a mink!"
> Ryan: "Minx!"

She was likable, funny, and forced Simon to explain himself, ruining his attempt to engage in flirtatious banter. While people may or may not have related to her lack of education, they probably responded positively to her forthright manner, buoyant optimism, self-confidence, and lack of fear. And according to this exchange she was thrifty too. Randy: "I like the shoes, though." Kellie perkily replied, "Thank you! I got them on sale!" Later in the same show, wearing a tight, slinky dress, she hopped up on a stool, saying, "I'll try not to bust the seam" (March 14, 2006).

Likability is an intangible, emotional component of a person's character. Candor and shared characteristics with audience members can promote likability. Being considered authentically ordinary is an essential component to being well-liked. Being well-liked can only help in a competition determined by audience vote. In the season-seven finale, Ryan described the two finalists as "two ordinary people achieve[ing] the extraordinary" (May 21, 2008). David Cook and David Archuleta were two ordinary people who managed to triumph over thousands of other people to reach the finale.

SONG CHOICE

For a contestant, the song-selection discussion can suggest authentic emotions and help build celebrity. The contestant can explain his or her motivation and reveal a memory connected to a song. This discussion also heightens the illusion that the audience sees much of what occurs backstage and has significant insight into contestants' characters. The pool of songs the contestants may choose from is unclear. Whatever the reason a contestant states for choosing a song, he or she must opt to sing a song that fits his or her archetype. A rocker singing a country song doesn't sound right; the judges do not hesitate to express that. As Ryan stated in season four, standing at a soundboard in the theatre, "Song selection is key if they want to stay in the competition and continue the whole fame-and-fortune thing" (May 3, 2005).

The right song choice is stressed again and again, in every season. In an example, Simon noted about Vonzell Solomon's choice in season four, "Because you have to make the audience want to vote for you. And with that performance, you want to pick up the phone. That's the difference of a good song" (April 12, 2005). Not only does a correct song selection make voters pick up phones, but also it can change viewers' perception of the contestant. Simon pointed that out in season six to contestant Gina Glockson:

> It was your best performance. This is all about knowing who you are, choosing the right song, the transformation from three or four weeks ago to tonight is literally chalk and cheese. It was not just your best performance, the best performance tonight. (March 27, 2007)

Song selection illustrates several aspects of authenticity: authenticity through backstage emotional revelations, authenticity by adding to archetype, talent as a component of authenticity, and authentic celebrity status. Each of these are explicated and substantiated below.

Song selection illustrates authenticity through backstage emotional disclosure

Emotional authenticity is fashioned in a variety of forums, including offstage interviews, packages about home life, and performances. Explaining how and why a contestant chose a certain song is another opportunity to create emotional authenticity. Contestants often use this moment to illuminate an aspect of their personalities through emotion.

Season-seven contestant Kristy Lee Cook stated, "I'm going to get emotionally connected with the song and give it my best" (April 14, 2008). She knew the importance of understanding the emotions in the song and expressing them, even if she was a little unsure which ones to underscore. Emotional connection is the primary reason contestants use to explain song choice, even if they can't eloquently explain the emotions connected to it. Season-seven contestant Chikizie debated his song choice with Byrd, the vocal coach, and members of the band: "But then Byrd, the voice coach, Matt, Rickey, talked to me and everything, said you know, you've got to follow your heart. So I paid attention for once and picked a song that means a lot to me" (March 25, 2008). He couldn't explain why it meant a lot, but he knew it contained emotion that spoke to him.

Other contestants were a little more articulate when explaining emotions associated with songs. Sitting at a soundboard, which cultivated an aura of music-industry credibility, season-four contestant Anthony Federov explained his reason for a song choice by noting,

> This song is about having faith in what you do, no matter how tough things get in life, you know. So that's why I'm singing 'I Knew You Were Waiting for Me' because I believe that in the end something really special is waiting for me." (March 22, 2005)

He exuded confidence, persevering to accomplish something, even if he was not sure what it was. Similarly, Vonzell Solomon claimed that "I picked this song 'cause it's going to show off my personality and I just want to have some fun this week" (May 3, 2005). She wanted to amuse the audience and entertain through an upbeat song that encourages good feelings.

In a more emotionally sentimental moment, fifth-season contestant Kevin Covais clarified:

> I chose this song because I think it's very fitting for me in particular. As a sixteen-year-old, I have yet to experience true love. And uh, I'll get a little sappy with ya, it's something I look forward to and it's something that I want to last forever when I get a hold of it. So, ya know, hopefully it happens for me. (March 21, 2006)

He sweetly revealed a romantic side to his character and a piece of emotional authenticity.

The right song selection can create a swelling of emotion in the audience, as well as in the contestant. Paula had tears in her eyes and on her face as she spoke to Elliot Yamin:

> You move me. You celebrate what this competition is all about and you know, I spent the day yesterday watching the tapes of when everyone first started and you've moved me from the beginning. (Her voice is breaking because she's crying.) But you are a handsome, evolved performer. (Simon is laughing next to her). You are an American idol. You are. (April 25, 2006)

The right song selection enables a seemingly emotionally authentic performance. The wrong song selection prohibits them from being expressed at all.

Song Selection Illustrates Authenticity by Confirming Archetype

It is important for each contestant to choose a song that displays vocal talent as well as fits with the archetype each represents, adding authenticity. As discussed earlier, while there may be small differences between a contestant and his or her archetypal predecessors, in general each contestant must by and large adhere to his or her archetype in order to convey authenticity. Therefore, choosing songs in accordance with the archetype is necessary. If the theme of the night is in conflict with a contestant's image, he or she must somehow alter the song to suit the idea of who he or she is.

The first two seasons exhibited many archetypes. By the third season, most contestants conformed to their *American Idol* predecessors' archetypes. For example, third-season contestant Amy Adams matched the girl-next-door archetype. Paula commented to her,

> You're so in the pocket of tonight's theme. You picked a song that is unbelievable and so true to who you are as we see you and you could release this song right now and it would be number one on the country charts. (March 23, 2004)

By picking the right song, Amy appeared "true to who you are as we see you." She may not be true to herself, but the song matches her *American Idol* image. Choosing the right song matches her to an archetype, yet can also differentiate her somewhat from those that came before her in the archetype.

Season four introduced a new male archetype: rock. While Ryan Starr and Nikki McKibbin originated the female version of the rocker in season one, Bo Bice began the male rocker in season four. Bo was the southern rocker, similar to the members of the seminal southern band Lynyrd Skynyrd. Subsequent versions included hard-rocking dad Chris Daughtry and emo rocker David Cook. But in season four, Bo was innovative, unusual, and different. He had to choose the right songs: either they needed to be rock songs or they needed to be sung in a rock manner. On the top-sixteen-male performance night, he sang, "I'll Be" by Edwin McCain, a country love song. Simon commented:

> I think it was absolutely, 100 percent the right kind of song for you. I think you'd have a hit record with that song again. And right now, at this stage in the competition Bo, I think it's your competition to lose. That's how good I think you're doing. I really do. (March 7, 2005)

"I'll Be" is a country song that has southern overtones, as most country songs do, and is a plaintive love song. Sung in Bo's raspy voice, it suited his image, proving that he didn't need to sing overtly "rock" songs, but that a rocker could be romantic and tender. The slow song added to his image, creating a more fleshed-out rock persona.

Season-five contestant Bucky Covington was very clearly a country singer. On March 28, 2006, he sang Tim McGraw's "Real Good Man," and Randy noted, "This is definitely the right kinda song for you. It's definite you're back at home in your country suit. I like that better, man." On the evening that the music of rock band Queen was featured, Paula claimed to Bucky "You made 'Fat Bottomed Girls' a country song, a country rock song" (April 11, 2006). First, Bucky demonstrated that his chosen genre of music suited him well. Next, he was able to twist a rock song into supporting his image. Both helped produce and round out his archetype, proving his authenticity as a country singer.

Another season-five contestant, Ace Young, had more difficulty. While a pop singer, he tried different styles on for size, not quite sticking with one image. On the top-ten show, he stated, "This week I'm going to show America a little more of a rock edge." After performing the rock song "Drops of Jupiter," Randy commented that it was "the wrong song for you and you didn't sing it well" (March 28, 2006). A rock song didn't add to his pop image and consequently, it didn't sound authentic. While it also may have been a poor song for him to sing vocally, choosing it muddled his archetype.

The season-five female country singer, Kellie Pickler, sang the Patsy Cline song "Walking After Midnight" during 1950s week. Kellie's southern twang made a country song seem even more country. Paula noticed this, saying it was a "true, authentic Kellie Pickler performance" (March 21, 2006). While Paula didn't detail exactly why it was an authentic performance, it can be assumed that it seemed authentic because it corresponded with Kellie's image. It was truthful because it fit the archetype Kellie represented, and thus Kellie herself.

Season-six rocker dad Chris Daughtry was excellent at modifying songs to fit his archetype. After singing the Red Hot Chili Peppers version of the Stevie Wonder song "Higher Ground," Randy noted: "Every week you figure out a way, no matter what goes on . . . to make it your own." Simon agreed, saying,

> You know I hate it when people don't take risks—they do a karaoke version of the song. It is so important that you choose the song well, which you did, you did make it your own, and it stands up, like I said, in the real world. That's what this show is all about. Best tonight, by a mile. (March 14, 2006)

Every song Chris sang corresponded to his rock ideal, including "Walk the Line" by Johnny Cash. Randy commented on this performance, saying,

> What I loved about this, that you took a song that everyone knows, you put a different spin to it, made it an alternative record that fit exactly who you are. And I love every week that you know exactly who you are and you never veer from that path. (March 21, 2006)

Daughtry matched a very recognizable and popular archetype. Song selection always helped his image.

Song selection can not only add to image, but also help form image. Season-six competitor Phil Stacy explained his song choice of "Blacktop Ends" by Keith Urban, for the top-seven performance, saying: "The reason that I picked the song is because it reminds me of growing up in a place like Kansas and really just getting to and playing in the dirt and being a country boy." After he performed, Randy commented: "Who would've thought, definitely not me. From an accomplished producer, you're going to have a career, man, in country music. That was really good. Who knew? I didn't know. Now I know Phil Stacy."

Paula liked it, saying, "It does seem like this is the genre you feel most comfortable in. . . . You sounded good from the beginning to the middle to the end."

And Simon agreed,

> Well, it's only taken ten weeks, Phil. I have to say I agree with what the guys said. This is the only time I heard you in a long, long time where I thought you chose a good song, I thought you sounded good and even more surprisingly, we actually saw some personality. (April 17, 2007)

Phil himself agreed with the judges. Phil explained, "This is the first night I went out and had fun. This is my genre; this is what I want to do" (April 17, 2007). By the top-seven performance night, Phil decided he was going to be true to himself and his hometown roots, and be a country singer. After this night, he attempted to sing the rest of his songs in a country fashion. His choice of the Keith Urban song highlighted his ultimate decision to conform to a believably authentic country archetype.

The next season, country singer Kristy Lee Cook attempted to adapt a Beatles' song into a country song. She explained,

> I'm singing "Eight Days a Week." I'm changing the song into a country song and I know Simon said he liked people taking risks. So I'm going to take this big risk and I'm just hoping and praying that he likes it and I'm able to impress him. (March 11, 2008)

Ryan then introduced her performance as: "'Eight Days a Week,' country style, from Kristy Lee Cook," reiterating her adaptation.

After the performance, Paula said, "I didn't like it and I'll tell you why. I just feel like you can't take what we say, go for the country thing, too much to heart. You've got be able to kind of infuse your own thing. I didn't get it."

Simon concurred:

> I thought it was horrendous, actually. You sounded like Dolly Parton on helium. Seriously. Sorry. Kristy, it was a very brave but probably foolish thing to do because that song just doesn't work in that style. It was like being at a ghastly country fair or something. Just a couple of banjo players and you. I didn't like it, sorry. (March 11, 2008)

Ryan interrupted, "Just so that we're clear, you gave her the advice to stay country, go country, and then you come back and say it was hideous."

Simon continued, "Yes, because the arrangement of the song was hideous, Ryan. There's one thing to have country style but to change the song like that was just wrong" (March 11, 2008).

The advice for Kristy to "go country" was a recommendation for her to clearly stick to a specific archetype. Kristy's intention in rearranging the Beatles' song was correct. She wanted to enhance her image by altering a pop song to fit her country archetype. Unfortunately, it backfired for a variety of reasons. Both the unusual arrangement and the strange experience of hearing a beloved song completely different probably affected the judges. This is a somewhat rare example of a disliked song revision. However, it was a unique performance that completely corresponded with her image.

Rocker Amanda Overmeyer did the same with the opposite reaction. Randy remarked after her performance, "You took a Beatles' song and took it to a southern club, a southern bar and rocked it out. I loved it, dude. I gotta give you props. Very cool. Very cool" (March 11, 2008). Amanda's version of "You Can't Do That" took a less well-known song and altered it just a little, mainly through vocal style, hardly actually changing the music the band played. Amanda's song was most likely better received than Kristy's because she changed a less popular song in a less dramatic fashion. Both songs matched the singers' archetypes, adding to their authenticity.

Later that season, another contestant struggled with archetype issues and song selection. After Jason Castro performed, Simon stated:

> We don't recognize you at the moment. For the last two weeks this is not the Jason we put in the competition. There was no attempt to make the arrangement your own. You struggled through both songs. They were both forgettable. I think you're going to look back on tonight and think I don't know who this person is. (April 29, 2008)

Simon's reaction is guidance for voters. The judging informed voters of the differences between Jason's recognized archetype and how he presented himself that night through song choice. Jason's song selections did not add to or correspond with his established singer-songwriter archetype. If song choices do not enhance an image, then they are a wasted chance to further develop the contestants' characters and augment authenticity.

Taylor Hicks always used songs to add to his blue-eyed-soul archetype. As Randy pronounced, "You don't care who sings it, you don't care what it's from, you don't care where it is, 'cause you always make it your own. That's the way to live it, baby. America, that's a hot one right there" (May 23, 2006). Taylor was continuously, authentically his archetype and used every song to add to authenticity.

Song Selection Illustrates Talent

Another component of authenticity is vocal talent. A contestant must truly be talented in order to be authentic on the show. Therefore, a crucial component of song selection is ensuring that the song choice flatters the person singing it. As Randy advised, "Always pick songs that can give 'em what you got" (February 27, 2008). The audience must believe a contestant's talent is genuine. While song choice can affect if a contestant makes it or not, talent cannot be faked, especially early in the audition process, when a contestant first must sing without accompaniment, and then with only a piano, or small band.

Contestants shared their thoughts about trying to sound good. Diana DeGarmo thought she picked a good song to "showcase them [the band] and myself" (April 27, 2004), while George Huff noted that he "tried so many songs in my rehearsal. I really want people to focus on my vocals" (March 9, 2004). Vonzell Solomon chose a song because: "It's very challenging and it's going to show off my range" (April 5, 2005).

The judges also realize the importance of song selection in establishing authentic talent. Simon said to Carly Smithson,

> I think you are an incredible singer. I don't think any of these girls can touch you vocally. But what you haven't done is chosen that one song that gives you the absolute moment and at the point you do, and I think you're going to be around for a long time, you're the girl they've all got to beat. (February 27, 2008)

Without the correct song, the audiences may not recognize talent and its corresponding authenticity.

Song Selection Illustrates Celebrity Status

The right song can help establish the singer as an authentic celebrity. The idea that a song could be a current chart hit illustrates contemporary fame. Additionally, fan adulation of a contestant singing a song also demonstrates authentic celebrity. During season seven, audience members could immediately purchase a contestant's version of a song on iTunes. The singles were not allowed to chart, which could arguably influence voting (voters may have been more likely to vote for a contestant who already seemed to be popular). Simply that the performances could be purchased augmented celebrity status. Being able to buy a contestant's song online placed them with famous peers who also had songs for sale on iTunes.

The judges emphasized celebrity status through song selection by commenting that the performance could be a contemporary hit, or that the contestant should make records like the song he or she just sang. If she should make albums like the song she just sang, then she is ready to be a celebrity.

In 2007, Blake Lewis sang the Maroon 5 song "This Love" and Randy noted, "That's the kind of record you should make. That's a good vibe for you" (May 15, 2007). Similarly, Randy commented about Melinda Doolittle singing Tina Turner's "Nutbush City Limits": "When you make your record you should put some of that in your repertoire. It's a good thing for you. It'll lengthen your career" (May 15, 2007). Again, Randy said to Brooke White about singing "Jolene" by Dolly Parton: "I definitely think that this kind of music suits your voice—you could make an album like this. It was all right" (April 1, 2008).

Paula also offered guidance, as she did to Kristy Lee Cook, "How the arrangement was done, I think you could have a hit with it in the country-western world" (April 14, 2008). The show rhetorically positioned the contestants as authentic celebrities, an idea that is enhanced through the correct song selection.

In another example, Randy made clear to David Cook: "I think more than almost anyone else on this show you are ready to record an album. You're a hot recording artist. I've never stood up once this season—that was the most brilliant performance yet (he stood up and applauded in a standing ovation)" (April 14, 2008). Song selection demonstrates talent, emotional authenticity, and enhances contestant archetypes, as well as illustrates celebrity status.

OVERALL AUTHENTICITY: EXAMPLES

Each season, the contestants must build their authenticity by crafting the right combination of talent, emotional candor, and familiar archetype. Each contestant tries to attain authenticity by revealing personal thoughts and emotions and choosing songs that fit his or her image that he or she can sing well.

Song-decision explanations add to contestants' apparent transparency and thus authenticity. Additionally, being able to explicitly demonstrate emotions is a key to being perceived as authentic. Visible emotions create a personality and allow the audience a way to personally connect with the contestants. As Paula said during the top-24 eliminations on the seventh season, "We look at personalities, we look at how you present yourself" (February 13, 2008). Paula was looking for personality as displayed through emotions. Following are significant case studies of authenticity constructions.

Season One

In season one, the competition structure was still forming, particularly the archetypes. Ability and distinctiveness were stressed, but different from what? That was never really answered. In the subsequent seasons, the answer is: different, yet still similar to those who came before you. In the first season, in a show untested by the American audience, the contestants had no previous seasons' season script to follow. While the show itself is modeled after *Pop Idol*, the American contestants seemed unsure how to present themselves and had the misfortune of having to demonstrate archetypes for the following contestants. While this season did feature offstage interviews and packages, the subsequent contestants had the luxury of more air time and longer seasons to fully establish authenticity, as well as the benefit of having seen the show before and knowing what to expect. The season-one contestants were flying without a net, unsure of what the audience wanted and uncertain about how to act.

While what the audience wanted was unclear, the judges were plain about desiring unique individuals who showed a lot of personality. Accordingly, the contestants tried to be different from each other. One aspect they all had in common was being very physically attractive and thin. The largest woman in the first top ten was Kelly Clarkson, who had an average build. There was also a great amount of racial diversity. Of the top ten, five contestants were white, one was of mixed race, three were black, and one was Hispanic. The first-season archetypes, discussed in chapter four in detail, include the two female rockers Ryan Starr and Nikki McKibbin.

Ryan Starr was eliminated from the show on July 31, 2002. Her departure highlight video included her voice-over saying, "An idol is somebody you look up to. I'm not scared to be who I am, on or off camera. I'm not trying to be this different person." She embraced the idea of being unique and herself at the same time. Similarly, Nikki said, "It's not an issue for me to try to be different, it's my thing" (August 13, 2002). The subsequent seasons emphasized more than originality; they also stressed authenticity and emotion.

While some season-one contestants presented emotions, the importance of emotional revelations in creating authenticity was not as perceptible as in later seasons.

Season Three: Battle of the Divas

On season three, authenticity was heavily emphasized. Talent, emotions, and archetypal coherence were all stressed. This season was known as the battle of the divas, for the high number of female rhythm-and-blues singers. Fantasia, LaToya London, and Jennifer Hudson competed, as did strong competitors Jasmine Trias and Diana DeGarmo. Fantasia, LaToya, Diana, and Jennifer struggled to differentiate themselves through emotional displays. This season featured many packages about home life before *American Idol* and stressed original, authentic personalities and sentiments.

Jennifer Hudson had trouble expressing her authenticity to the judges. Paula explained it by saying:

> First of all, I love your hair like that. You look great. I also feel that, Jennifer, I want to get your personality back again. Like I feel like you're kinda holding back a little bit. I don't know quite how to articulate it, but I don't feel like the real Jennifer is punching through. I don't know quite what it is. I don't know. Just not sure. (March 23, 2004)

Jennifer's response was sassy: "Well, I got soul, I got style, I got attitude, I can give you whatever you want." Paula and Randy liked the enthusiasm in her response. Paula said "I wanna see more of that!" while Randy agreed, "That's it right there!"

She later improved with the Elton John song "The Circle of Life." Paula observed, "Now I hope you realize that you get it. This is who you are, Jennifer. Don't go off the road. Stay on that right path. This is who you are" (April 6, 2004). The following week, Paula happily continued,

> Jennifer, nothing pleases me more than to see you shine. Now you're really, really coming into your own and I see that this is truly who you are and you're feelin' it. You look beautiful tonight. The only thing I felt was wrong is that I didn't get to hear you sing more. (April 14, 2004)

Unfortunately, by the next week Jennifer may have felt too comfortable, saying, "How do I think I'm doing? I feel like I'm getting to be me now. This is Jen's world" (April 20, 2004). She was voted off that week in a surprise elimination.

LaToya London had similar issues to Jennifer. Randy fawned about her vocal skill:

> You're the best that I've heard every week since we started. Oh my god. Wow. Passion, singing, control, performance, star power, everything. I mean this is what it's about. This is what the search for a superstar is about to me. That was brilliant. (February 24, 2004)

However, she was reserved and had to struggle to overcome what appeared to be a reticent nature.

Simon noted, "I think at some stage you are going to have to learn to get your personality over. Because trust me, this competition is as much personality as it is vocal ability. But I think you did really well tonight" (March 23, 2004). Simon warned her that emotional authenticity is as necessary as vocal talent. Simon brought this point up again much later in the competition:

> My only criticism about you, Latoya, is we've known you about ten, eleven weeks. I know nothing more about you now, in terms of your personality, than I did ten weeks ago. And disco was one week where you could at least try something different 'cause I've seen and heard that before from you. (May 11, 2004)

Simon criticized her lack of emotions and therefore personality, since her vocal ability was unquestionable. If LaToya had built more emotional authenticity, she may have won season three.

Runner-up Diana DeGarmo had an ongoing authenticity problem that Simon pointed out week after week. According to Simon, Diana was more like a singing mannequin than an emotive performer. At her audition, Diana sang a cute song, displayed a huge, teeth-baring grin, performed little-girl dance moves, and was dressed entirely in pink and black. She explained that she wore only pink and black because they were her favorite colors. She had an innocent image, full of hope, yet also artificial and contrived. Her performance was very clearly premeditated. From her smile to her pink beret, everything appeared carefully planned and deliberated. Simon disliked her young, cute image. His problem at the audition was her "horrible routine, the cutesy routine you do with your song." (January 28, 2004). To Simon, Diana was simply an excellent vocal technician. He explained to Diana:

> My only slight worry is when I see those images of you as a child. It's like a product of the pop farm. . . . You're missing the point. It's very effortless for you this. But you're not to me yet connecting with the audience properly. (March 23, 2004)

Simon repeated himself the following week, saying, "Vocally, I think you are very good. However, visually, it was rather like a high school production. In other words, you could see that in any school across America. And that's what we're looking for, which is star quality. I just don't get it, Diana."

Randy disputed Simon's claim: "But you can't hear that in any school across America. That's the difference. And it's a singing show" (March 30, 2004). Randy focused on her voice while Simon was concerned with the overall performance. The next week Simon chose not to comment on her performance and the following week, when the top eight performed songs from movies, Paula called the performance a "nice job, but not the most moving performance" while Simon said "Your problem is at the moment, is that I don't think the audience gets who you are, 'cause you are like an overgrown child . . . it was just automatic and predictable. Sorry" (April 14, 2004). According to Simon, she performed without any emotion imbuing the song. Without emotion, Diana didn't come across as genuine. However, she improved over the next few weeks and received mediocre praise from Simon. Twice, he gave her a "7 out of 10. That's a compliment" (May 18, 2004).

Fantasia had talent and an engaging personality deemed authentic by the judges. Latoya London, Diana DeGarmo, Jennifer Hudson, and Fantasia were all exceptional singers. However, the judges pointed out Fantasia's authenticity, both in emotions and talent, from the beginning of the season. In the semifinal round, after her performance, Randy complimented her: "This is what this competition is all about, uniqueness, unique vocal sound, attitude. You look like an old pro up there. Give yourself a round; that was good" (February 10, 2004).

Simon agreed, remarking,

> Fantasia, let me put this into context. Every other person sitting in the room before you needed this competition to get a recording contract. The difference is, you actually wouldn't even need this competition to become what I think you're destined to be, which is a star. You have exactly what this competition needs: you're original, you're quirky, you have talent, you're different. You're just a star, congratulations. (February 10, 2004)

Fantasia's personality was encapsulated in her happy-go-lucky charm, ebullient dancing, and wide smile. She continually exuded joy. This, combined with her musical selections, created her image of a happy rhythm-and-blues singer meets pop performer.

Throughout her tenure on the show, the judges adored her. Simon praised her authenticity on March 30, 2004, saying, "Fantasia, you are the real thing."

He restated his pleasure with her on April 20, 2004, noting: "What you've done is show your personality and thank god you're still in this competition. Really, you did great."

Randy agreed, observing that "There's a star in your voice, it's that tone, it's about tone people" (May 4, 2004).

Fantasia explained to Ryan that she tries to incorporate herself into every song she performs, saying on May 3, 2004, "I try to do every song (Fan)tasia way."

Fantasia was different from the other competitors because she expressed her emotions, injecting them into every interview and performance. She had great emotional authenticity, an outstanding voice, and fit into the established rhythm-and-blues archetype, yet showed enough differences to those similar to her to be considered unique. On the episode where the top-three contestants performed, Paula said:

> Fantasia, you know what this competition is about. It's being unique; it's capturing the spirit, tapping into the hearts of millions of people. You came here, you came to win, you celebrate each and every week, everyone's having a blast. You picked the right songs, good for you. (May 18, 2004)

Simon continued the analysis of her performance,

> You demonstrated there, I think what we're looking for in this competition, which is the difference between copying someone badly or originality. And that was, to me, you demonstrated what you're all about. And what hopefully what we're looking for, maybe you found in this competition, which is somebody who is original. It was superb, well done. (May 18, 2004)

Fantasia epitomized the *American Idol* authentic celebrity narrative. She demonstrated authenticity composed of talent and emotions within a constructed atmosphere. She also corresponded to an archetypal image with small differences from archetypal predecessors. Fantasia was ordinary though her emotions and extraordinary through her talent. She was an authentic celebrity. Fantasia won season three and continued on to fame and success, both on Broadway and on the music charts.

Season Five: Chris Daughtry and Taylor Hicks

Like season three, season five also contained intriguing examples of the importance of authenticity. Two contestants on season five were unique for their authenticity: Chris Daughtry and Taylor Hicks. Both were consistent in representing who they were and garnered respect. Both also were emotionally demonstrative.

Early on, Chris was lauded for his emotive performances. In a semifinal round, Simon commented, "Chris, this was the only performance—remember this guys (he pointed to the other contestants)—which stands up in the real world. It was the only memorable, good, real performance I saw tonight" (March 1, 2006). Simon recognized the performance as emotionally authentic and also true to Chris's archetype.

The following week, Randy noted about Chris: "I love people that know who they are, know where their voice is. I think that's really important when you do a show like this. That's one of the things I'm definitely looking for" (March 8, 2006). Chris consistently projected the same version of himself.

Guest judge Barry Manilow also noticed the strength in Chris's personality: "Out of all them, Chris seems to have a real good handle on and who he is and what is his strength" (March 21, 2006). Chris shaped his performances around his chosen archetype. Paula discerned this as well, saying "You are so true to who you are and you don't ever abandon that. That's what's so amazing about you Chris. And it's like you grow and you grow each week." Simon had the highest praise: "I think you are the first artist we've ever had on this show who's actually refused to compromise and for that, a round of applause" (March 21, 2006). However, even when steadfast, he was able to be flexible. Paula noted, "You showed versatility and you still stayed true to who you are," (April 4, 2006).

The show also presented Taylor Hicks as authentic through displayed emotions, interviews, packages, and performances. At his initial audition, Taylor proclaimed allegiance to himself: "I'm not trying to be somebody different. I'm not having somebody trying to make me look any different. I'm just who I am" (January 31, 2006). His claim that he wouldn't and couldn't change himself was believable. Over the course of the season, Simon repeatedly denigrated Taylor and his dancing, but Taylor seemed to make no effort to change his stage persona to suit what Simon thought was appropriate. He remained authentic to his blue-eyed-soul, exuberant, and likable persona. The judges could like or dislike him, but he didn't change. Whereas Chris was commended for staying true to his archetype, Taylor was derided for establishing a new one on the show.

On the top-four performance evening of season five, the contestants sang the music of Elvis Presley. Host Ryan Seacrest asked Taylor how he felt about performing his second song "In the Ghetto." Taylor replied, "I'm ecstatic about it. I've always wanted to sing this number and it's one of the more soulful numbers that Elvis did" (May 9, 2006). Taylor identified as a blue-eyed-soul singer and consistently chose rocking soul songs, in order to better to show off his exuberant personality and unusual spastic dancing. After Taylor performed "Jailhouse Rock" the same night as "In the Ghetto,"

Paula noted: "You're original as ever," while Simon commented, "The dancing was hideous." The dancing distracted Simon, but it added to Taylor's emotional authenticity and originality.

Season Seven: Amanda Overmeyer, Brooke White, Robbie Carrico, and David Cook

Season seven featured several contestants who appeared remarkably authentic. Randy pointed this out early in the season when Ryan asked him about the new group of contestants: "This season I saw more originality from these dudes up here, man. So hopefully they're going to show us some of that. Keep it real, dudes." Simon concurred and stated, "I'm looking for personality, originality, and obviously, you've got to sing well" (February 19, 2008). Keeping those precursors in mind, comparisons of Amanda Overmeyer to Brooke White and Robbie Carrico to David Cook are illuminating.

Amanda Overmeyer solidly fit the rocker archetype, but changed it a little by being a nurse. She was the rock nurse. She said, "People have embraced me as being rock. My goal when I go out there is just to solidify the presence of rock and roll, I guess." Amanda dressed the part as well. Paula said,

> I just love everything that you do. I do—because you're authentic. You don't separate. It doesn't matter what people say, is she a one-trick pony. You're the real deal. This is who you are and I love every bit of it. (February 20, 2008)

Simon agreed:

> I gotta tell you, Amanda, I really like you. I like you because I genuinely think you're authentic. You stand out in a crowd. It's not the best performance . . . I hope you stay around for a while. I think at some point you have to come out and prove you're a great singer again, 'cause you are. (February 20, 2008)

The judges were in accord. Amanda authentically conformed to the rock archetype.

But the next week, the judges did not find her as authentic. She performed "Carry On Wayward Son" by the rock group Kansas. However, before her performance, a package revealed that she was not as hard-edged as she seemed. In fact, she was a bookworm and loved to read. Simon said,

> I thought actually in your film you came over as very natural, very cool. And then everything felt contrived, from what you wear, from this to terrible hair, from the indulgent song. Everything. None of it felt natural or real, it was like you're in your own little world with quite an ugly song. And I couldn't wait for

> it to finish. I really, really didn't get it. If you want the popular vote, you're not going to get it by doing that. (February 27, 2008)

The details in the package did not fit her archetype. Her onstage persona was deflated by her contradictory backstage emotional reveal. Her emotionally authentic revelations did not fit her tough archetype. Typically, she was shown riding a motorcycle, not reading a book about the music industry. Her authenticity faltered.

In contrast, Brooke White began and ended the season as authentic. She illustrated a new *American Idol* archetype, which is difficult, but was helped by the new season-seven rule that allowed contestants to play their own instruments. Brooke established the female singer-songwriter archetype, even though she didn't write her own songs on the show. However, her delivery was so believable and familiar (greatly aided by her ability to play piano and guitar) that often it appeared as if she could have potentially written the songs she performed.

Paula championed her throughout the season. Paula noted to Brooke, "What this competition is about is originality. And that's what I love about you. You have your thing. That's what this whole business is about, you hear someone for the first time and you identify that spark" (February 20, 2008). Two months later Paula felt the same. "Every ounce of you is totally authentic to who you are. And that's a beautiful thing—you're identifiable" (April 14, 2008). Brooke remained loyal to her archetype and also was emotionally authentic, often crying on camera and voicing her insecurities. Her only detriment was her lack of variety; her performances were very similar to each other, which denotes consistency, but also was monotonous. Nevertheless, she was emotionally authentic offstage and on, which was more than Amanda could deliver.

Robbie Carrico appeared as the archetypal rocker: long hair, bandana, jeans, chain wallet, and wailing voice. He truly looked the part. However, he had also toured in a pop band, opening for Britney Spears. He was a rocker with a pop past. Could such a thing exist? Robbie claimed to be authentic, saying, "Tonight I want to show them me. Just put it all out on stage and leave my soul right out there for the audience to see" (February 19, 2008). He performed a rock version of "One" by Harry Nilsson, and Paula believed his rock archetype, "You're authentic. You really are. You stay true to who you are. You look fantastic tonight. And you know what, the vocals weren't over the top, they were in the pocket, as they should be."

Simon liked him but was unsure: "I'm struggling a little bit with what Paula said about the authenticity. I'm not quite sure yet whether you're very comfortable in this rock thing. Is it really you or are you a pop singer? You tell me."

> Robbie: "No, definitely not a pop singer. I can sing anything but I love rock and this is me."

Ryan: "Why do you question the authenticity?"
Simon: "I don't know. I question it because I'm not convinced yet, Ryan."
Ryan: "Do you buy it, Randy, Paula?"
Paula: "I buy it, since day one."
Randy: "He's a rocker."
Ryan: "Dressed like a rocker, but kind of looks like Justin Timberlake, doesn't he?" Robbie shook his head no.

The problem still existed the following week. Before the performance, Ryan asked Robbie: "Robbie, Simon has argued a little bit of the authenticity factor with your performances. What do you need to do tonight to prove that it really is you?"

Robbie replied:

> You know, I get up there every time and I be me. I am me. I guess he has to understand that there are so many different levels of rock, so many different styles and all. I guess he's just focused on one and doesn't think that I'm the one that's in his mind. So, get used to it. (February 26, 2008)

After he performed "Hot Blooded" by Foreigner, the judges debated again. Randy changed his mind from last week, explaining:

> I must say I agree with Simon. I don't know if the rock thing is really, really your thing, 'cause I mean, that's like one of my favorite songs, you know. Foreigner, he has an amazing voice. It doesn't have enough uhhhh to really be like a rock voice. Nothing really came out. It didn't quite go there, it was just okay. (February 26, 2008)

Paula defended Robbie, saying, "I bet it could drive you crazy having someone say it's not authentic, it's not who you are. How can anyone know who you are but you?"

Randy interrupted Paula, "The thing is, dude, rock is like an attitude, not something you should prove."

Simon, seeing his point from last week had taken hold, backed off: "Robbie, you don't have to get so defensive about it. It's only an opinion. I don't know you well, as Paula said. Actually, I thought the vocal was okay tonight" (February 26, 2008).

Robbie was voted off that week. Ryan asked if Simon had any advice for Robbie, and Simon replied, "Robbie, I've said this from day one—it's authenticity, and I think the public saw what we saw, that was the problem. It just never ever felt real. Sorry" (February 28, 2008). Could Robbie be a rocker just because he dressed the part and claimed to be? Could he be rock when he toured with Britney Spears in 2000? Did his background, which he freely admitted to, preclude him from authenticity? Apparently, his pop past did preclude his authenticity as a rocker. Simon (and the popular press outside of the show) planted a seed of doubt. Robbie was never believed as the rock archetype due to his affiliation with pop bands and Britney Spears. Authenticity must resound with your background, appearance, and emotions. Rock and Britney Spears didn't mix. Ironically, while on a show initially dedicated to creating pop stars, Robbie's pop star past hurt him. He lacked authenticity, a problem that ultimately destroyed his chance.

In contrast with Robbie Carrico, David Cook was perceived to be authentic. David Cook had a much more believably authentic rocker past—he truly did struggle in a band and worked as a bartender to support his dream. Whereas Robbie looked as if he was wearing a 1980s hair-band "rocker" costume, David wore contemporary styles and clothes. David looked as if his clothing was not a costume he donned to audition for a role, but what he actually wore on a daily basis. He also played electric guitar, adding to his authenticity as a musician. Randy commented,

> So listen, man, you got the Marshall [amplifier], you got the Les Paul [guitar]. For me man, to me, you're the real rocker of the boys this year. I actually really, really liked it; it was a smart choice of song. It was cool. I really kind of believe you. (February 26, 2008)

Paula concurred, "David, you are the real deal. You got it, you got it, you got it. It was very smart, very fun, very relevant to who you are. I loved it" (February 26, 2008).

As the weeks continued, David continued to rock, emoting through his performances as well as offstage. David adapted many songs to his genre, including several ballads and an Andrew Lloyd Webber Broadway song. He rocked on stage to The Beatles on March 11, 2008, when Randy averred, "You can definitely rock on *Idol*, that proves it right there. You can rock out on *Idol*; you can rock out on The Beatles. You're doing your thing."

Paula loved the performance: "I've been telling everyone you are the dark horse in this. You're a thoroughbred. We all know you're the front man with your band, but you are a front man here."

And even Simon liked David's performances consistently throughout the season, saying, "David, I thought it was brilliant. If this show remains a talent competition rather than a popularity competition, you actually could win this entire show."

When not performing, David was open with his emotions and articulate when describing them. When asked which *Idol* moment was his most memorable, he responded,

> So many memorable moments. If there was one I had to pick I would say last Tuesday. It was like the epitome of what I wanted to do when I started playing music. It's to have that rock concert feel, with the lights and the smoke and the audience. It was intense. I remember the very euphoric feeling, like nothing could hurt me. (March 18, 2008)

As a self-proclaimed "wordnerd" who loved crossword puzzles, David was communicative and a good speaker, facilitating emotional authenticity. While this was not a feature that many rockers would admit to, for David it worked. He embraced the intellectual rocker mantle, the thinking man's (or woman's) idol. Rock does not preclude intelligence, if that was always an aspect of the contestant's personality. Being bright was inherently a part of David's persona, as opposed to Amanda's image, which focused more on riding a motorcycle and being tough. David was never a tough guy, but rather a sensitive, intelligent, and self-proclaimed rocker nerd, an image that tapped into a relatable emotion.

The judges raved about David's moving performance of Chris Cornell's restrained version of Michael Jackson's "Billy Jean." Randy admiringly said,

> David Cook. I gotta say, I think you're probably the most original, the most bold contestant we've ever had. And I'm telling you, that performance right there tonight, on that joint you just did—you might be the one to win the whole lot. Blazing, molten hot! (March 25, 2008)

Randy recognized the creativity and inventiveness in David's performances as well as the courage necessary to perform something different and risky. Paula said, "I gotta tell you, how smart you are, how brave you are, and how willing you are to stretch the boundaries. You go right to the edge without going over. I think you're brilliant." Paula identified David's qualities: intelligent, daring, enthusiastic, and prepared. Simon continued the theme: "David, that was brave, it could have either been insane or amazing, and I have to tell you—it was amazing" (March 25, 2008). The judges appreciated David's emotional sensitivity, excellent voice, and willingness to try something innovative. Not only was he the season's rocker, he was talented, emotional, and authentically believable. David won the competition.

CONCLUSION

American Idol helps engender celebrities through rhetorical devices and manipulation of image, as well as through exposure of seemingly truthful emotions, and archetypal coherence. Authenticity, displayed through emotional revelations in interviews and packages, talent, backstage access, and song selection, appears to be a factor of popularity. The winners of *American Idol* are arguably depicted as extremely authentic on the program, more so than their competitors. This may be for a combination of reasons: amount of time on the program, ease of fitting into an archetype, talent, and proclivity to emote. Kelly Clarkson, Ruben Studdard, Fantasia, Carrie Underwood, Taylor Hicks, Jordin Sparks, and David Cook were all represented as very authentic: their emotions were laid bare to the audience, and their backstage personas and song selections all agreed with their archetypal narratives. Of course, all of them also displayed vocal talent, an essential component to authenticity within *American Idol*. They came across as authentic, or true to who they were, so that they were also familiar and accessible. They seemed friendly and real, yet worthy of fame through their authenticity. Perception of authenticity, on *American Idol* and arguably elsewhere, is thus a necessary condition for celebrity.

The judges repeatedly urge the contestants to be themselves, show who they are, choose the right songs to express themselves, and to be unique. This builds authenticity, and in turn, celebrity. On stage, the contestants attempt to imbue emotions into their performances. However, these performances are just that—performances. The contestants act out emotions in order to construct authenticity. The judges stress that choosing the right song is necessary to establish contestants as unique (and to show off vocal skill) even as most contestants emulate those who came before them in their chosen musical genre. However, authenticity is built not just on stage, but also in premade packages, interviews, and behind-the-scenes footage.

The audience at home glimpses backstage frequently. It appears that the contestants work hard, adding merit to their talent. The contestants hang around, prep for the show, and most importantly, prepare for their performances. The audience can see contestants rehearse, and work with vocal coaches and the band. Audiences also witness the typically backstage audition process, an often grueling and humiliating experience. Through the backstage scenes, the contestants are positioned as ordinary people, working hard to create good performances. The audience is also made to feel as a part of the show. Backstage access permits emotional demonstrations and gives a seemingly truthful perspective of the contestants, adding a crucial component of celebrity creation on *American Idol*.

As audiences seem to accept what the show provides, proven through the high ratings of the show, the contestants morph from ordinary people into extraordinary celebrities. Acceptance is clearly illustrated through voting and high television ratings. While audiences may be aware of inherent falsity of reality television (after all, a talent competition of this grandiose nature does not occur outside of television) it can be presumed the savvy audience does not care (Andrejevic, 2004). The contestants and the show seek to provide the audience with moments of authenticity (Hill, 2005) to reveal their true selves through emotions. Authenticity helps diminish the distance between a viewer and a contestant, while the vocal skill and celebrity increases the distance between the viewer and the contestant. Thus, authenticity simultaneously fosters both a relationship and distance, helping create celebrity. Emotions, when judged as real, impart authenticity, a necessary requirement to achieving fame in today's false world of celebrity. Celebrity may be fleeting and constructed, but the person in it must resound as truthful through his or her emotions.

Chapter 6

Narratives of Transformation to Celebrity

While the *American Idol* contestants must be authentic, another component of the show is the confirmation of celebrity status. Within *American Idol,* a large portion of the show is given over to the verification, maintenance, and promotion of current and future idols. By nature, celebrities are "the few, known by the many" (Evans, 2005a, p. 1). Famous guests frequently perform, or mentor and critique the contestants. Even dead celebrities occasionally make an appearance, such as Elvis singing a duet with Celine Dion on the season-six *Idol Gives Back* special. With such an emphasis on celebrity, the show creates a foregone conclusion that being a contestant on the show will make him or her a well-known personality (even if the viewers know better). On the very first episode, Simon Cowell detailed the fame stakes of the show:

> In England the winner and the runner-up became the most famous people in the United Kingdom. They went on to sell more records in one week than any other artist has ever done in the history of pop music. If the same thing happens, the stakes are so high. The public is the people who get to choose their idol. (June 11, 2002)

The winner of *American Idol* potentially can be a huge star, earning millions of dollars. Even those participants who do not win are considered luminaries while on the show. This chapter explores how the show constructs celebrity.

AMERICAN IDOL AND CELEBRITY

American Idol contestants are already considered to be celebrities and deserving of fame because of their talent and hard work. The show reiterates this fame through the overall fame discourse, including discussion of the contestants as stars, contestants participating in events such as photo shoots and red carpet premieres, association with current celebrities, image changes, and fan adulation. Additionally, performing on television weekly for thirty million people adds to fame. According to Gamson (1994) there are two common narratives explaining how and why people achieve fame: merit and construction. *American Idol* utilizes both of these narratives, combining them into a singular narrative.

CELEBRITY EVIDENCE

On the first season, Ryan Seacrest claimed the winner would "live a celebrity life with fame and fortune" (July 23, 2002). These and other repetitious phrases cement the idea that the winner will become renowned, if he or she isn't already by the end of the show. Not only is the winner of *American Idol* seemingly guaranteed fame, but also the contestants are treated as famous throughout the run of the show.

Most celebrities walk red carpets, wear designer clothes, are the center of photo shoots, appear in magazines, and are often beloved and besieged by autograph-seeking fans. By emphasizing these aspects of fame, *American Idol* creates the notion that the contestants are already famous when their names are in fact just beginning to be known. Thus, the normally consequential results of fame create the illusion of fame, which in turn actually becomes fame for a short time.

CELEBRITY DISCOURSE: VERBAL AND VISUAL

Throughout each season and each episode of the show, verbal and visual discourse positions the contestants as celebrities. In an evocative example, the logo itself, a cursive version of the words *American Idol*, has semiotic connotations. The A in the *American Idol* logo is shaped like a star, indicative of the show's star-making promise.

There are aspects of singing onstage or on television that cannot help but be associated with celebrity. Large spotlights and klieg lights are reminiscent of Hollywood glamour, fame, and star power. The massive stage itself reminds audiences of the idea that the world is the stage for these contestants. As Ryan explained "It's not just a stage—this is a platform, a platform that will launch someone into superstardom" (February 19, 2008).

The celebrity discourse begins with the setting of the show itself: the stage. The stage represents many different ideas, including fame and success.

Beginning in the initial audition episodes, the prospective contestants perform on a stage. By Hollywood week, each contestant performs in the center of a stage, holds a microphone, and typically stands under a spotlight while surrounded by flashing lights and swooping cameras. Interspersed in the performance are rapid reaction shots of the adoring audience. The contestant is positioned as a star onstage. A frequently used camera shot has the camera very far from the stage in the rafters of the theatre, creating celebrity by adding distance between the TV audience and the performer. The camera pulls back all the way to the ceiling, emphasizing how sizable the room is, how famous the performer is and may continue to be, and just how high the stakes are on *American Idol.* Moreover, merely singing and performing with a full band on a stage of this size, with an audience, provides a modicum of status. Even in the early stages of each season, when the audience consists of other auditioners, the contestants are still positioned as potential celebrities. They perform on a large stage, with the judges on a raised platform in front of the stage, akin to how the rest of the show is blocked, including the big finale. Auditioning and performing on the stage is different than singing in a small room in front of three people and a camera crew—the setting emphasizes the true significance of the show and its career and fame-launching capabilities. Auditioning on the stage is a reminder of how the next twelve weeks could be—performing live, demonstrating vocal aptitude, displaying charisma, and cultivating fame. As Simon said during the season-seven Hollywood week, "This stage—it makes you or breaks you. If you're comfortable up there, then it's a really, really good sign" (February 12, 2008). Season-seven contestant Amanda Overmeyer realized the symbolism of singing on the big stage and the amount of celebrity at risk. She commented: "My most memorable moment thus far was playing on the big stage. Hell of a lot better than the flatbed trucks I'm used to playing on" (March 18, 2008).

Each season, the stage changes a little. The yearly changes are negligible as all the stages have an outsized production in common—an abundance of lights, cameras, a substantial audience, and a significant number of television screens in the production. Each season the number of screens on the stage increases as the number of contestants dwindles. By the 2008 seventh-season finale, there were upwards of twenty-five television screens in various sizes on the stage. Screens imply celebrity status by having contestants' faces on them, hinting at what is to come in the future (being on TV and being even more famous), and are representative of all of the people watching on home televisions. Being on television triggers and bolsters fame. The celebrity possibilities of the competition are made more apparent through the use of screens—the reason the contestants are singing is to keep appearing on screens and to be on even more screens throughout their careers. In season seven, while each contestant spoke with Ryan before the voting results were revealed, a picture of him or her singing the night before filled a colossal screen behind the real contestant. It was literally larger than life, bigger, better and more beautiful than the average person, making the contestant more famous than his or her earlier incarnation as an average individual. The oversized screens both establish the contestant as a celebrity and suggest a superior fame to come. The enormous screens help reinforce and create celebrity status.

CONFIRMATION OF FAME THROUGH MEDIA FRENZY

Contestants' celebrity status is also confirmed through the show discourse and association with the show itself. An exchange between contestant Latoya London and Paula highlighted the celebrity discourse apparent on the program. Paula asked Latoya, "How does it feel being on this new stage in front of a live audience?" Latoya replied, "It feels great; I feel like a star right now." (May 11, 2004). However, Latoya was eliminated the next night. She came in fourth on season three and in her final highlight reel said:

> I got to experience how it feels to be famous—cameras around you, pictures, and autograph signings and just all the things that you dream about doing as an entertainer I've gotten to experience through *American Idol*. To perform for thirty million people is just like a dream come true. *American Idol* made it happen instantly for me. I just got a chance for America to see me and that's what I've always wanted. I'll never ever forget this experience and I'm so lucky to be here right now. (May 12, 2004)

Latoya, very humble and self-aware, realized that the show helped to confirm her fame. Her fame wasn't earned solely through talent, charisma, or star quality, all of which she had. *American Idol*'s confirmation of this fame, through cameras, photographs, and autograph signings, was also necessary.

A part of the show's celebrity discourse is demonstrating how famous the contestants have become. *American Idol* presents a media frenzy surrounding the current contestants. Every season, the contestants attend movie premieres, award shows, and like events, where they walk the red carpet, are photographed, and sign autographs. The *American Idol* camera crew is there to record and broadcast all of it. As first-season cohost Brian Dunkleman pointed out, "The camera doesn't lie, these contestants are ready for the limelight." Justin Guarini agreed, "We've all dreamt of photo shoots, of appearances, of doing interviews. It's just mind blowing that we're finally here" (July 31, 2002). An example of such an appearance occurred in the first season when the contestants presented an award at the MTV Video Music Awards. Ryan introduced the packaged segment reporting their appearance by saying, "Here is a taste of the media frenzy that has surrounded these guys" (September 3, 2002). Through the program's positioning of the contestants as already famous, the contestants tautologically become seemingly more famous.

In every season, similar celebrity packages are utilized repeatedly. For example, on the February 21, 2008 episode the top twenty-four contestants were shown arriving in Hollywood and attending a photo shoot. The audience had a behind-the-scenes view at the photo shoot, including being able to see the cameras and lighting apparatus. This allows the audience insight into the contestants' celebrity lifestyle. Next, the dressed-up contestants exited limousines onto a red carpet, where paparazzi shouted at them and took photographs. After that, another photo shoot took place with the contestants posing together under bright lights. The music playing during this montage was by season-five contestant Chris Daughtry, linking his success and fame with the current contestants. Discourse about contestants attending events as celebrities, with all the attendant trappings, position them as celebrities. As season-four contestant Vonzell Solomon said, "One thing I love about *American Idol* is we get to do photo shoots, we get to dress up, get our hair done, get our makeup done and be beautiful" (May 3, 2005). Once they appear as celebrities, ipso facto, they are celebrities.

Anthony Federov, a season-four contestant, said, sitting in a dressing room in front of a makeup mirror:

> The first time I realized *American Idol* is kind of like the real deal is when I set my foot on the red carpet for the first time. And you know, that whole night I was flying in the clouds and I was just so blown away by all that attention I was getting. And you know at that moment I realized that like my life is taking on a whole new meaning. (May 3, 2005)

Anthony was being interviewed while being prepped for something that required he be made up. This visual depiction alone created the idea that he was an actor or singer, someone who performs in public or for media. Media and *American Idol* confirmed him as a celebrity. He was one of the few known by the many.

I'M FAMOUS

The judges, the contestants, and the hosts verbally utter the celebrity discourse on the show. On the first season of *American Idol*, before anyone knew the scope of fame that the show could bring in America, contestant Nikki McKibbin said, "No matter how famous I am, I'm always, always, going to come back home" (August 27, 2002). She spoke as if she was already famous and would always be famous, claiming that she would never forget her hometown roots. She assumed, and thus the audience could assume, that fame is a guarantee from the show. In 2008, she was a contestant again—this time on *Celebrity Rehab*. In 2002, perhaps Nikki felt she was a celebrity because of the discourse of the show as spoken through the hosts. For example, in one show cohost Brian Dunkleman said that the "finalists were adjusting to their new found fame" and that "in less than four months, the singers had gone from virtual unknowns to household names." The contestants echoed this: R.J. Helton said that he was "recognized on the street" and Kelly Clarkson thought, "signing autographs is weird" (August 13, 2002). The show consistently has the judges, hosts, and the contestants themselves establishing and reacting to fame.

First-season cohost Brian Dunkleman continued the repeated referral of the contestants as celebrities by saying, "You made them household names, now let's hear you make some noise for your final five" (August 14, 2002). At the same time, the show paradoxically claimed that they could be even more famous. Simon commented after Kelly Clarkson performed: "I truly believe you will be a huge star at the end of this show" (August 14, 2002).

Similarly, contestants and their families contend they are celebrities while on the show, and their discourse throughout the show tautologically confirms this. On season three, contestant Camile Velasco's stepfather, James, was interviewed about her: "A month ago she was a waitress at IHOP. And now she's doing ET and Ryan Seacrest. She has no idea what a big celebrity she's become" (March 16, 2004). The more Camile appeared in the press, the bigger a celebrity she became, and the more she appeared in the press. And so, cyclically, her celebrity and her amount of press grew.

Other aspects of fame are money, power, and prestige. Implicitly, contestant Phil Stacy referred to these benefits when Ryan asked him about missing his kids while on the show. Phil answered: "Even though it's hard to be away from them, we know in the long run, it'll make a better life for them." Why would Phil's children have a better life if their father was on *American Idol*? As Ryan said, "'Cause daddy's famous. Daddy's becoming famous" (April 24, 2007). Seventh-season contestant Chikizie reiterated, "It's really insane, I mean now people know my name" (March 12, 2008).

CELEBRITY DISCOURSE: CURRENT CELEBRITIES

Well-known celebrities appearing on the show lend their credibility, image, and connotative qualities to the show and the contestants. They also gain publicity from their show appearences. Merely standing on the *American*

Idol stage places a contestant in the company of Celine Dion, Jennifer Lopez, Mariah Carey, Tony Bennett, and other performers. These known stars also attract audience viewers. Throughout most of the *American Idol* seasons (excluding season one) guest judges and mentors frequently appeared. Even celebrities sitting in the audience, such as Harry Connick Jr., Jenny McCarthy, and Eva Longoria, silently endorse the contestants and the show with their appearance and support. Clearly, the show and the contestants must be significant if celebrities enjoying watching.

Celebrity judges, accomplished musicians, songwriters, and record industry insiders lend credibility to the show and to the whole competition—as they take it seriously, it must truly have value and credible results. More importantly, guest celebrities indicate the potential future of the contestants. Guests have included performer Verdeen White (from Earth, Wind, and Fire), songwriters like Diane Warren (hits for Cher, Aerosmith, and LeAnn Rimes, among others) and Lamont Dozier (hits for Martha & the Vandellas, The Supremes, and The Four Tops, among others), and music executive Clive Davis. Guests are identifiable as celebrities by the title of celebrity judge or mentor. A video package also explains their successes. Many celebrity guests, such as Lionel Richie, Smoky Robinson, Lamont Dozier, Gladys Knight, and Neil Diamond, are from the old-guard music industry. Other celebrity judges and mentors, such as Mariah Carey, Dolly Parton, Gwen Stefani, and Jennifer Lopez, are top-tier music industry stars. The celebrity guests link *American Idol* to music history and contextualize it within the history of music. Through observing and referencing the lineage of music history and guest judges' places within it, *American Idol* situates its own place in the annals of music. The guest judges perform a range of tasks: legitimizing the competition with their presence, providing a place for *American Idol* in music history, and sharing their celebrity status with the contestants on the show.

As the show grew more popular over the years, the status of the guest celebrities grew as well. The second season included guests little known to younger audiences, such as Donna Summer and Neil Sedaka. In later seasons, the celebrity guests included Martina McBride and Jon Bon Jovi, singers with current Billboard hits and projects to publicize, while still including established older stars such as Tony Bennett and Barry Manilow. The contestants are shown interacting with the celebrities to some extent and also sing the reputable celebrities' songs. The show suggests the illustrious guests and contestants also have a master/apprentice relationship, as the term *guest-mentor* (used in later seasons) implies.

Celebrities on the show act as judges and mentors, teaching and often nicely critiquing contestants' performances of their songs. They are generally shown as encouraging, usually finding something nice to say about everyone. These celebrity guests demonstrate what the contestants aspire to be, successful musicians and celebrities, as well as show the level of celebrity that they may attain. Williamson (1978), in her seminal book on decoding advertisements, showed that advertisements work by shifting signified concepts from one signifier to another. Following that line of thought, celebrity status and credibility can be shifted from a bona fide celebrity to a contestant through association. For example, Diana DeGarmo noted that, "It was really cool to meet Ms. Estefan, because she just has this presence" (April 27, 2004). Being able to meet and work with Gloria Estefan, who most likely wouldn't waste her time on an untalented noncelebrity, confirmed the celebrity status of the contestants.

The celebrity judges also act as experts, offering their thoughts on the contestants, lending authority and trustworthiness to the contestants' talents. The established stars position the contestants as celebrities for the audience. For example, Elton John said about Fantasia (who sang one of his songs), "I'd give her a record contract now. Blew me out of the water. I think that's probably the best that song's ever sounded, me included. And I'm not saying that lightly. I thought she was incredible" (April 6, 2004). Clearly, Elton John has musical expertise, particularly about his own songs. His praise gave Fantasia's talent validity. He also praised Jennifer Hudson, saying, "Jennifer Hudson blew me away. She sent chills up my spine. It was my favorite performance out of the whole lot. Just that voice is astonishing" (April 6, 2004). Likewise, Dolly Parton and Jon Bon Jovi are credible sources and their expertise validates contestants' skill. As David Cook said, "To have Dolly offer me any compliment, let alone a compliment about my voice, is very awe-inspiring" (April 1, 2008). Season-five contestant Ace Young was awestruck by Rod Stewart's validation: "Having a legend comment on your voice is a trip. He gave me a stamp of approval and that's the most fulfilling thing I could have ever asked for" (April 18, 2006).

The performance songs popularity is exemplified by The Beatles songs performed in season seven. The Beatles, of course, are arguably the most popular band of all time, with almost all of their songs being instantly recognizable. The show informed the audience of the musical significance of The Beatles with a package about their very impressive musical history. Having the contestants sing the music of The Beatles, while certainly challenging in various ways, also linked them to The Beatles and their enormous success.

Even if a contestant and a celebrity did not interact, qualities of a well-known musician were often used to describe a contestant. Typically, the judges claimed a contestant looks like, sounds like, or simply reminds them of an established star. This validates that the contestants are stars. Season-two contestant Trenyce reminded Paula of "a brand new Diana Ross" (March 11, 2003). On season three, Camile was reminiscent of Lauryn Hill to Randy

(February 17, 2003), Diana DeGarmo caused Simon to think of when "Christina Aguilera was nice" (February 10, 2003), and Fantasia was "like a young Aretha," expressed Randy (March 4, 2003). Latoya London reminded Paula of a young Gladys Knight and Simon of Tina Turner (March 23, 2004). Melinda Doolittle again made Simon think of Tina Turner (May 1, 2007), while Randy deemed Jordin Sparks to be like Beyonce (May 15, 2007). These comparisons give the audience markers to judge the contestants by and a way to understand them. Associating Fantasia with Aretha Franklin provided the audience with a large set of mostly favorable preconceived ideas.

Even when the comments are negative, being associated with a celebrity still isn't bad. The contestants still have merit through the comparisons. Judging Nikko Smith's performance on the season-five top-twenty show, Simon said "One thousand percent improvement. My only downside is, you're looking like Bobby Brown" (March 1, 2006). Bobby Brown, while maybe not the best singer to unknowingly (or even fully aware) emulate, still has a modicum of celebrity and did have a musical career. On season three, Simon said that contestant Amy Adams reminded him of Jay Leno.

> Amy exclaimed, "That's not nice!"
> Simon replied, "His sister I mean."
> Amy asked, "Does Jay Leno have a sister?"
> Simon retorted: "He does now." (March 23, 2004)

This unfortunate exchange compared a funny contestant with a prominent chin to Jay Leno. While certainly not very nice, it was a memorable contrast. Amy Adams was now connected with Jay Leno. She may not have liked or wanted the Jay Leno comparison, but it probably helped her. After all, Jay Leno is the well-liked host of *The Tonight Show* and welcomed into millions of homes nightly. In fact, Jay heard about the comparison, mentioned it on TV, and then had Amy on his show. She was voted off tenth and was the first contestant not in the top five to appear on *The Tonight Show*. Simon's superficially negative association actually helped Amy.

Even the show judges, Simon, Paula, and Randy (later Kara DioGuardi and Ellen DeGeneres) are celebrities, and associating with them added to the celebrity status of contestants. Since the audience ultimately votes and decides the competition, a primary function of the judges is adding celebrity and viewers. As Sanjaya rhetorically asked backstage during the sixth season: "Dancing with Paula Abdul, how much better can it get?" (April 4, 2007). Of the judges, Paula began the program with the most celebrity due to her career as a pop singer. Arguably, all of the judges now have a degree of fame due to the show. At the initial auditions, many potential contestants profess their anxiety at having to perform in front of the judges. They may be nervous about being judged, but also may be somewhat star-struck at interacting with celebrities. Their reactions to Paula, Randy, and Simon add to the judges' overall conception as celebrities. The judges, in turn, lend some of their celebrity standing and credibility to the contestants. Their celebrity also gives them credibility as arbiters of talent.

CELEBRITY DISCOURSE: IMAGE, APPEARANCE, AND LIFE TRANSFORMATION

The celebrity image discourse of *American Idol* has two layers. The initial, most obvious layer focuses on being yourself and the importance of a strong personality, no matter appearance. The second, more subtle layer stresses that the acceptable image is that of conventional beauty: slim build, full hair, physically sexy, good-looking face, and attractively dressed no matter which archetype the contestant embodies. While Randy and Paula often deplored Simon's negative comments about appearance, it cannot be denied that he spoke of common beauty stereotypes and what he thought would succeed in the record industry. The show promotes stereotypical beauty ideals and transparently works to transform contestants into the accepted physical and sartorial model of a star. Once contestants begin participating in the show, as Ryan stated, explaining a first season photo shoot for *Us Weekly*, they get "the star treatment. From makeup experts, to super hair stylists and even all new wardrobes" (July 31, 2002). Image and appearance are extremely significant within the celebrity discourse. While the contestants already deserve celebrity due to talent and hard work, their image transformation confirms this celebrity.

IMAGE TRANSFORMATION

The overriding discourse theme of *American Idol* pertaining to image is that looks are unimportant and the contestants are judged on their voices alone. However, just beneath that initial layer of democratic competition is the secondary layer stressing contemporary beauty norms. The secondary beauty emphasis is demonstrated through comments about attractiveness (disguised as concern about image) and weight. *American Idol's* emphasis on beauty ideals becomes clear when compared to an orchestra audition. Prospective orchestra members audition behind a screen, allowing judges to concentrate solely on the musical skill. *American Idol* offers no such protection. Often contestants attempt to conform to beauty standards by transforming from their initial image to an accepted ideal.

Throughout the seasons of the show, no one says out loud that a contestant is not attractive enough to be on the show or win. Rather, image is a code word for attractiveness. Image is composed of dress, weight, and overall appearance. Also, over the course of a season, contestants are often shown changing to a clarified, simplistic image that fits one of the previously discussed archetypes. They are also depicted striving to lose weight or gain muscle, dress better, and appear more attractive. In all these ways, contestants endeavor to fit the stereotypical image of a music star. Appearances have the possibility to make or break a contestant, as Simon pointed out to Jason Castro in season seven: "Where you're lucky is that this a TV show, not a radio show, because your face sold that" (March 18, 2008). Jason's good looks triumphed over his occasional mediocre performances and saved him from possible elimination.

IMAGE: ATTRACTIVENESS

American Idol asserts that contestants are natural-born celebrities due to skill; the show simply celebrates their abilities and helps them blossom into their rightful positions as celebrities. *American Idol* corroborates contestants' celebrity through photo shoots and red-carpet events, but still emphasizes that they are likable, relatable people, who just happen to have talent. Over the course of each season, contestants often modify their images. Discourse on the show transparently stresses this transformation. The audience can see the transformation of images through spoken discourse and visuals of the contestants picking out new clothes, etc. On the second season, Ryan explained, "We can help with their image, help with performance, but once they're out there, they're all on their own out there on that stage" (February 4, 2003). The show coaches the contestants on how to be a celebrity, but inherent talent or star quality is necessary. On the same show, Simon reiterated the importance of image: "I work for a record label and I'm telling you, image is important" (February 4, 2003). Even celebrity judges reinforce the importance of having a good image. Lionel Richie commented to Kimberly Caldwell in season two: "The look is getting better and better, you're getting stronger on that" (April 8, 2003).

Scott Savol's season-four audition displayed the tension between the supposed democratic competition only judging vocal quality and the significant concern about appearance in a show that mimics the world of celebrity. Scott Savol was a bespectacled overweight white man, had a buzzed haircut, had small features in a large face, and was not conventionally physically attractive.

Scott Savol appeared in front of the judges on February 1, 2005, for his initial audition and said, "I'm here to become the next American idol."

Simon asked, "What've you got?"

Scott replied, "I got the voice, I got the look, I got the confidence, I got the personality."

Simon's response: "I'll give you three out of four on that." After listening to Scott demonstrate his vocal talent, Simon said:

> This is where this competition for somebody like you is a necessity. Because, *A*, you wouldn't get a foot in a record company's door under normal circumstances, and if I'm being honest with you, if you walked into a record company, if you got in, I think they'd throw you out even before you sang. And that's the reality. Which is why this show's a good thing. Because I think you have a good voice. I just don't think you're an American idol. That's my problem. (February 1, 2005)

Paula said, "I didn't expect to hear a voice like that come out of you. Which is always pleasantly surprising."

Celebrity guest judge LL Cool J agreed and said:

> They would never see you coming. You know what. So much in this country is based on looks and image and blah blah blah, but you know what? I would sure love to give America a chance to decide for themselves with you. 'Cause I think you got a lot of talent. (February 1, 2005)

After deciding on Scott's fate, Randy yelled, "Welcome to Hollywood baby!" As Scott grabbed his yellow ticket to Hollywood and walked out of the door, Ryan said in a voice-over "Success and validation."

Simon implied that while the program doesn't discriminate because of weight or physical features, the public and the music industry do. The show is a reflection of the music industry in particular and the celebrity world at large. To be a successful pop singer, beauty or stereotypical attractiveness is a necessity.

The judges again discussed Scott Savol during the top-twelve male performance show. After Scott performed, Simon said, "It's going to be interesting what the audience makes of you. You know I said it when we first met, you don't look like an American idol. But you know that. But you have a nice voice."

Randy asked, "What does that mean though? C'mon."

Simon replied, "Well, he doesn't. You work for a record company, you know what I mean."

Randy rejoined, "I mean, but did Ruben, did Clay? Neither one of them looked like American idols either."

Simon said, "Well, we put him this far, I'm glad you're here. It was okay, it was good, it wasn't fantastic."

Ryan jumped onstage and said, "It's okay, some say Simon doesn't look like he should be on television either. Remember, America votes" (February 21, 2005). America may have decided the American idol, however, Simon was the voice of the music industry.

In the above exchange, Simon seemingly didn't want to say that Scott was not attractive enough to be the American idol. Randy challenged Simon, but Simon did not want to engage in a debate about what the idol should look like. Simon explicitly called on the music industry as a deciding factor of what the American idol should look like, but Randy, overweight himself, refused to acknowledge accepted ideals of attractiveness and the role they may play in voting. Even if the music industry and society at large support stereotyped beauty and sex appeal ideals, often the *American Idol* voting audience doesn't seem to care. Scott Savol came in fifth place in the fourth season, despite the fact that he didn't fit conventional beauty ideals.

Similarly, when season-five winner Taylor Hicks auditioned, Simon again didn't think he was *American Idol* material because he did not fit accepted standards. Taylor Hicks had prematurely gray hair and a stocky physique, and he was not conventionally handsome. Simon said immediately after Taylor Hicks auditioned: "My problem is, and I've always said this, it is not just about the voice, and you prove that."

Paula said to Taylor: "You've got a good personality (Simon looked at her in disbelief). You're just like a good energy, good character, just good performer."

> Simon: "I disagree completely."
> Randy: "Naw, c'mon."
> Simon: "Completely and utterly."
> Randy: "You don't think this guy could be commercial?"
> Simon: "Nope."
> Randy: "Why?"
> Simon: "Because this is a guy who should be singing backgrounds, not in the spotlight."
> Randy: "Backgrounds moving like that, what?"
> Paula: "I disagree."
> Simon: "They will not put you in the final group to be judged by the public." (January 31, 2006)

Simon was wrong. The voters didn't seem to care that Taylor didn't fit conventional attitudes about attractiveness while his closest competitor, Katherine McPhee, was a standard beauty. Of course, Simon was right in one sense. Neither Scott Savol nor Taylor Hicks have gone on to great commercial success.

Over the course of a season, many men and women adjust their looks to conform into an ideal image. Clay Aiken is the most obvious example of the transformation contestants undergo. Clay transformed from a classic nerd to an approachable and still somewhat geeky pop star. At Clay's January 28, 2003, audition, he sang for Randy and Simon (Paula was absent that day). Clay was painfully thin, wore glasses and a striped button-down shirt, loose khaki pants, and had short, dark brown hair. He looked like the consummate stereotype of a nerd, missing only the pocket protector. After Clay sang, Simon commented, "Ok, very good. You don't look like a pop star, but you've got a great voice."

Randy agreed, "Yeah, it's weird and wild. I say yes. You can work on your style; you got a really good voice though. It's just really wild for me to hear that voice coming out of this . . ."

Clay interrupted, "Little white boy?"

Randy finished, "He can sing."

Simon said, "Welcome to Hollywood."

Ryan ended the scene by saying, "So Clay looked wrong, but sounded right."

Clay didn't fit the typical pop-star mold, but had the vocal chops. Over the course of the season, particularly in the first month or so, Clay shifted his image to that of a pop star, but retained a little of the accessible, awkward boy-next-door look. For Hollywood week, Clay ditched the glasses and wore a somewhat hipper outfit of jeans, a necklace, and a blue button-down over a gray t-shirt. Simon said, "Clay, you don't look like a typical pop star. But you are unique" (January 29, 2003). Clay's image gradually progressed, altering a little from the quintessential nerd.

During the semi-final round, Ryan interviewed Clay and asked, "Now, we first met you, you looked a little differently. Talking about the image and the look, what have you changed about yourself?"

Clay responded,

> Well, I had people help me out in Atlanta (where he was living), had some people pick stuff for me. I didn't like that. So I just thought this time, I'm not gonna listen to anybody else, I'm just gonna do what I want to do. (February 11, 2003)

He continued,

> I may not have the boyband look, or the pop-star look, but I'm comfortable with myself and I'm comfortable with my style. This is who I am. So what if the judges don't like it—hopefully you guys do (to the camera/audience). (February 11, 2003)

Clay had authentic talent and celebrity; he needed to work on his image. And even as Clay's image changed, he was comfortable with it and continued to be himself. He blamed the nerd look on others who had helped him choose an outfit, but even so, his general image changed from extremely nerdy to only somewhat nerdy. By the wild-card show on March 4, 2003, Clay admitted, "Image is an important factor but I'm always going to have these (points to ears)."

After he performed, Randy commented to him, "You got your look together, did your homework, yay dude." Wisely, Clay kept his normal guy-next-door aura that appealed to fans.

Although Clay managed to inflect some pop star into his image of approachable geek, by April 15, 2003, Simon still said to Clay, "I like you better with my eyes closed."

On *American Idol Rewind* (a program that is a repeat of a previously aired show combined with supplementary contestant commentary), in 2007, Clay recalled the season finale that starred himself and eventual winner Ruben Studdard. He said:

> As we were at Universal that night, preparing to go onstage, Ruben looked at me, and he always had this sage wisdom about him. He's a man of very few words but when he says something it's something that you remember. And he grabbed me and he pulled me close to him and he said, 'Who would have thought that two of us would be here' because this show has always been about who looks the best and who looks the part and who has the image to do the show. And he said, 'look, the fat boy and the skinny boy are on the finale of the show.' (April 26, 2007)

After Clay Aiken transformed on season two from a full-fledged geek into an approachable pop singer, future contestants followed in his footsteps. These contestants included glasses-wearing, thin, and nerdy season-four competitor Anthony Federov. On the top-ten performance episode, Anthony wore jeans and a long-sleeve, green button shirt, with no glasses. Randy remarked, "Innocent Anthony's kind of grown up now, took the glasses off, the whole nine. It was all right. It was definitely all right" (March 29, 2005). Later, on the top-six performance episode, Ryan complimented Anthony, saying "He's been hitting the gym twice a day, look at those arms."

Anthony reacted, "Thanks for embarrassing me."

Ryan rejoined, "I didn't embarrass you. The chicks love that" (April 26, 2005). Anthony was apparently trying to alter his physical image in an effort to appeal to more fans. He succeeded to an extent, finishing the competition in fourth place.

The next season, contestant Elliot Yamin began the season at his audition looking like an urban hoodlum, wearing sunglasses and baggy clothes, with a short-buzzed haircut and a goatee that exacerbated his overlapping and jumbled teeth. Throughout the course of the season, he appeared with a better haircut and more fashionable clothes, including suits and ties, which made him appear respectable and handsome. After the show ended, Elliot had his teeth fixed as well. On the top-four performance night, Ryan commented to Elliot, "Yamin, you have really evolved in this competition."

Elliot concurred, "I have. I've come a long way, man."

Ryan inquired, "Where's the guy we met a few months ago?"

Elliot replied, "He's history man, he's gone" (May 9, 2006). Elliot's initial unclear, blue-collar image was vanquished, replaced by the crooning, handsome, celebrity image.

Season-two contestant J.D. Adams epitomized the image that the show, and Simon in particular, values. J.D. Adams was a tall, conventionally attractive, blonde Caucasian man with a muscular build. He also claimed to be a descendent of Presidents John Adams and John Quincy Adams. He resembled Ken, Barbie's boyfriend, with an aristocratic pedigree. At his Los Angeles audition Simon said to him, "You are the business. You are what this competition is looking for. Seriously, what it's all about. Great-looking guy, nice personality, great voice" (January 21, 2003). J.D. was the total image package, but ultimately lacked the talent and inherent celebrity needed to succeed on the show. He was not one of the season-two top-twelve contestants.

Season two was the first season where image truly competed with vocals. Vocals won, as the top three contestants were Kimberly Locke, an overweight woman, Clay Aiken, an underweight nerd, and Ruben Studdard, an overweight man. All three illustrated previously infrequently depicted images for successful competitors on any television show. By the end of the season, on the finale pitting Ruben Studdard and Clay Aiken against each other, Simon voiced this idea by saying, "It's image versus talent. Talent won, America got it right" (May 20, 2003). Nevertheless, while talent may triumph over image when left up to the voters, the show's discourse continues to favor stereotypical attractive images.

IMAGE: WEIGHT

In general, discourse on *American Idol* claimed not to care about looks and image, but the secondary discourse of the show belied this claim. Even though the show paid lip service to the idea that overweight people can succeed on *American Idol* (and indeed, overweight contestants Ruben Studdard and Jordin Sparks did win), Simon's discourse in particular illustrates that he clearly favors stereotypically attractive people. Unattractive men typically proceeded further in the competition than unattractive women. Even Ruben said during the semi-final round, "I really didn't think I was the *American Idol* type" (February 11, 2003). It can be assumed he was obliquely referencing his un-pop-star-like physique. However, Ruben was never disparaged for his weight. Numerous female contestants, on the other hand, were frequently chided for a few extra pounds. Kimberly Locke, Vanessa Olivarez, Frenchie, and Mandisa, among others, were told to lose weight or put down for their size by Simon.

Overweight contestants, perhaps used to being treated poorly in their normal lives, often were insecure about their weight. On the very first episode, a very overweight African American woman named Jacquette Williams auditioned. Simon said to her, "Did you expect to get through today?"

She answered, "I don't want to say no, but no."

Simon asked, "Why not?"

Jacquette replied, "No, because I'm a big girl."

Simon relished telling her, "But you are through to the next round" (June 11, 2002). The show seemed to pride itself on not discriminating between overweight and average people.

However, the rest of the evidence points to a great concern with image and appearance, particularly noticeable in Simon's comments; he had no qualms calling a contestant overweight. As Simon commented to season-two contestant Vanessa Olivarez,

> I'm being truthful. This isn't going to make me very popular but I'm going to say it. What I would say to you—I would say, great voice, great personality, brilliant choice of song. I think you need to lose a few pounds. (February 18, 2003)

Vanessa retorted, "I have junk in my trunk, like J. Lo, and you love it, you just won't admit it."

Simon answered back, "I would just work out a bit more, I'm not being rude, just follow my advice" (February 18, 2003). If the judges do not like a contestant's image, they repeatedly comment on it.

The night after telling Vanessa to lose a few pounds, a recap of the season-two auditions aired. That episode included the audition of an average-sized woman named Lisa Leuschner, who in the following season made it to the semi-final round. Simon said to Lisa, "I think your voice is sensational. Really, really good. But you've got to get your act together. You also need to lose weight."

Randy: "Huh?"

Paula: "No, she doesn't."

Simon: "I'm only saying to you what I would be saying to an artist if they were signed to my record label. I say it to guys on my label. When they get overweight, I say lose weight! 'Cause this is an image business."

Paula: "I hope we keep perpetuating eating disorders."

Simon: "This is simply to do with what a record label would say to an artist" (February 19, 2003).

The discourse implied, judging by who was told to lose weight, that a woman's image requires a thin figure.

Also competing on season two were Frenchie Davis and Kimberly Locke, two big women with big voices to match. They were aware that they were not the norm on the show, particularly as season one featured no overweight men or women, and they were determined to succeed despite this perceived drawback. During Hollywood week the two paired up for the group competition. Frenchie said, "We're the two big girls. We have to sell it" (January 29, 2003). Later in the show, Kimberly Locke spoke with the stylist about what she should wear, saying "I got put in the fat girl category, so black" is the color she's going to wear. It was unclear whether she assigned herself to that category, or a producer on the show did. Kimberly Locke finished third on season two of *American Idol*. She later appeared on the 2007 season of *Celebrity Fit Club*, eventually losing forty pounds.

The most well-known weight comments made on the show by Simon concerned season-five contestant Mandisa. Mandisa was an obese African American woman. After Mandisa successfully auditioned in Chicago and left the room, the judges exchanged comments.

Simon: "Could we have a bigger stage this year?"

Paula: "She's got like a Frenchie growl."

Simon: "Forget Frenchie, she's like France" (January 17, 2006).

Upon learning she had made the top twenty-four, Mandisa confronted Simon, saying:

> Simon, a lot of people want me to say a lot of things to you. But this is what I want to say to you is that, yes, you hurt me, and I cried and it was painful. It really was, but I want you to know that I've forgiven you. And that you don't need someone to apologize in order to forgive somebody and I figure that if Jesus could die so that all of my wrongs could be forgiven, I can certainly extend that same grace to you. (February 15, 2006)

Simon responded, "Mandisa, I am humbled. Come here and give me a kiss," and they shared a hug. He never did apologize on air. Superficially, the show accepted Mandisa, extra pounds and all. However, more frequently than for other contestants, Mandisa was shot from the shoulders up, particularly during interview segments and with guest mentor Stevie Wonder. The only time the audience saw her entire body standing up was during performances.

Apparently women only up to a certain size are acceptable on the program. Mandisa was by far the largest woman to appear on the show during the first seven seasons, and the show attempted to minimize her size through camera shots and editing. Thus, the show itself endorsed a weight ideal through production. Mandisa was the ninth-place finalist.

The only male contestant who was ever told to lose weight on camera was season-two competitor Josh Gracin. On the top-twelve performance show Simon said to Josh, "I think you could lose a few pounds." Taken aback, Josh challenged Simon to "come up and do some pushups with me, we'll see how many you can hang for" (March 11, 2003). Josh competed the same season as Ruben, who was clearly more heavily built than Josh, yet no one ever told Ruben to lose weight on-air. This discrepancy implies that being heavy is acceptable for an African American man, yet not for a Caucasian man. While Ruben's nickname, "The Velvet Teddy Bear" implied he was teddy-bear-like, huggable and squeezable, Josh was overweight. On the other hand, Josh's image as a Marine might have made the extra weight more offensive than on Ruben. Perhaps Simon was more comfortable telling another white man to lose weight. At the same time, the verbal and visual *American Idol* celebrity discourse makes clear that extra pounds on women, regardless of race, are undesirable.

Notably, both Katherine McPhee and Carrie Underwood began their respective seasons about twenty pounds heavier than when their seasons ended. Both were beautiful women who started the season with pleasantly plump, average bodies and by the finale were quite thin pop stars. While surely the stress and pressure of performing weekly on national television may have had something to do with losing weight, the show's discursive emphasis on appearance may also have had something to do with this weight loss. Attractiveness is essential to celebrity.

SARTORIAL, HAIR, AND MAKEUP CHANGES

Contestants have two image goals: to adhere to a chosen archetype and to appear as an attractive celebrity. Each contestant may begin the show as a typical person, but ultimately attempts to transform into a more beautiful version of him or herself. The *American Idol* discourse asserts that clothing, hair, and makeup are often key components in changing an average person's image into one of a celebrity. Additionally, these components help contestants match archetypes. For instance, if a contestant declares himself or herself to be a country singer, he or she must retain that image through all style choices. Often, images constructed through clothing evolve as each contestant realizes, or decides, which archetypal role he or she fits into.

Contestants start the season appearing as normal, unfamous people and the farther they progress in the competition, the more their overall appearance resembles that of a celebrity and fits with his or her archetype. This coincides with Biressi and Nunn's (2005) contention that the sartorial aspects of fame are necessary to establish celebrity, particularly for celebrities who begin their rise to fame from working-class circumstances. While the contestants are already celebrities due to talent and star quality, their style needs to confirm this. Style and dress substantiate contestants' celebrity. Accordingly, by the end of season six Jordin Sparks was shown selecting a finale outfit with the couture designers Mark Badgley and James Mischka, an experience only the extremely wealthy or famous enjoy.

In 2007, season-two contestant Kimberly Locke said on *American Idol Rewind*:

> Being on *American Idol* probably gave me a sense of style and made me more aware of it and very conscious of it. You learn what looks good on you and also that everything you wear doesn't always translate on television. (April 12, 2007)

She continued, "It's interesting to watch yourself evolve into all of that because you want to keep your own personality in there but you also want to look like a pop star" (April 12, 2007).

Influence on Contestants

Occasionally contestants comment on the *American Idol* staff's influence on their appearance. Celebrification is a trial-and-error process. Bucky Covington, a singer within the country archetype, typically performed wearing jeans and boots. Early in the season, Ryan asserted, "There's nothing Hollywood about Bucky. We'll see how you change. Soon your hair will be spiked" (February 22, 2006). One particular night, in contrast to his general appearance, Bucky's long blond hair was brushed out, flowing and fluffy, more apt for an Osmond than Bucky. Simon commented, "I've got to talk about this Jessica Simpson hairstyle. I mean seriously, that has got to go!"

Bucky responded, "Totally not my idea."

Simon: "Obviously not" (March 14, 2006).

Contestants often seem to enjoy aspects of image construction. Scott Savol claimed that one of "the cool things about being on *American Idol*—I used to dress myself, now I have a fashion coordinator who puts all my outfits together that make me look good on TV, to make me more appealing to people" (May 3, 2005). Once, Ryan asked season-seven contestant Ramiele Malubay to name her favorite part of being on the show. Her answer was: "Hair and makeup. And these shoes are awesome" (February 20, 2008).

Season-five contestant Paris Bennett agreed with the joy of having a wide variety of sartorial choices, saying,

> The best thing I love about Hollywood lifestyle is I actually get to be me. I love to keep change; 'cause change allows you to see who you are. And fashion allows you to bring out your personality. And that's what I do because you can tell who I am through my fashion. (February 28, 2006)

Later in the season, Paris compared song choice with fashion choice, saying, "Every song allows me to show another side of me. And each hairstyle I have is a different side of me" (April 4, 2006).

Other contestants had style difficulty. After Simon labeled Latoya London "mumsy," she noted her problems with appearance. She said, "At the Los Angeles auditions, I didn't know what they were looking for. I didn't want to look mumsy, so after meeting with wardrobe I had to rethink" (February 24, 2004).

The judges often comment if appearance construction works or not. On season two, Kimberly Locke usually wore her hair naturally curly. One week, she straightened her hair. Simon said that while she had weird hair before, "you look good now" (April 22, 2003) with straight hair. This comment had racial connotations to it. Simon apparently disliked kinky African American hair and instead preferred the straight locks that often mark Caucasian beauty.

African American Latoya London also tried curly hair with extensions, instead of her normally close-cropped waves, and Randy and Simon immediately commented.

Randy said, "I'm not really quite feeling the hair."

Simon concurred, "I do agree with Randy, however, about the hair. It looks like you have a cat on your head. You look so much better with your normal hair."

Latoya, with grace and humor, replied, "I think it's a beautiful cat" (March 30, 2004). The next week, her hair was back to normal and the judges approved. The show prefers the kind of beauty that American media generally promote: thin, Caucasian, and young. The preference for youth is visible in the age limits of the show; a contestant cannot be older than 28 at the initial audition.

Sometimes an image problem contaminates an entire performance. Simon commented on Chikezie's performance in the semi-finals round of season seven, saying, "Here's my problem: I absolutely hated the entire performance and I'll tell you why. The suit is hideous." He continued after Chikezie's protestations, "The suit was hideous, the wink was hideous, the woo was hideous. It was all old-fashioned, corny, cheesy. To be honest with you this could have been something we filmed 40 years ago" (February 19, 2008). The clothing choice affected the entire image and thus the performance's reception.

Sometimes the stylists working for the show also ruin performances. Their effect shouldn't be obvious—the look needs to authentically fit the contestant. Simon remarked to George Huff,

> My only advice I'm gonna give you George is that at the moment, you are looking as if you are totally in the hands of the stylists. And I think you have to create now—it's what Ruben had last year, he had his own sense of style. And I think it'll be nice for you to develop that as well. (March 23, 2004)

George was wearing a nondescript light brown leather jacket with a brown button-down shirt and black pants. He looked nice, but not noticeable. George blamed his bland style on the stylist.

> Ryan: "Hey George, let's talk about your style, fashion for a moment. 'Cause it has evolved a bit, right?"
> George: "Yes, it has evolved. If you look at some old pictures of myself, you will see that I went through a whole bunch of different changes, you know. I probably will go through some more."
> Ryan: "Are you picking this out, or is there some help?"
> George: "I leave it all up to the stylist, 'cause I don't know what to do. I say, stylist, you know put me in something nice and this is their fault. It's all their fault!"
> Ryan: "Careful."
> George: "I like this, though."
> Ryan: "It looks good, man." (March 23, 2004)

As evidenced, the stylists and wardrobe staff have some influence over the appearance of the contestants. Contestants want to remain on their good side.

Early in the third season, Simon reiterated the importance of the stylist to Jennifer Hudson, whose memorable outfits included a shiny silver jumpsuit. He commented to Jennifer that she would have to "find a stylist who can see" (March 9, 2004). The discourse of the show implies that few contestants, if any, are able to create their own look. They rely on the knowledge and insight of the show's stylists to create an appearance suitable for a star that also conforms to their desired image.

Simon stressed the importance of celebrity image appearance again in season seven to Carly Smithson, who performed wearing an unremarkable outfit of tight red pants tucked into black boots, with an unflattering loose and long black top. He said,

> I also think, Carly, that you ought to have a word with whoever's dressing you at the moment. I think at this stage, without being rude, you've got to start looking more like a star. And I'm not seeing that progression at the moment. (April 1, 2008)

Carly already had the talent; she needed the image of a star to confirm her celebrity position. Paula countered this opinion, saying, "I think you're looking more beautiful and beautiful and evolving" (April 1, 2008). While Paula and Simon disagreed over Carly's appearance, the underlying message is the importance of image. The contestant should evolve over the season into a fully formed archetypal image. Most importantly, the contestant's image must be that of a star.

While the judges are often critical, they note when an image is positively working as well. They frequently declare that a contestant looks beautiful, lovely, or attractive. For example, on the season-seven top-three-contestant performance evening, Paula called Syesha "absolutely stunning" and Simon nonchalantly said she looked "gorgeous, by the way" (May 13, 2008). They frequently encouragingly commented on appearance. Randy observed to season-three competitor Jon Peter Lewis, "The good things: I like the coat, I like the shoes, I like the whole outfit. Like your energy, you got the personality; tonight the voice let you down" (March 30, 2004). Two weeks later, Paula disagreed, saying to Jon Peter Lewis, "I still feel though you're trying to find your style in the competition. I'm not quite sure that you've really locked into what your style is" (April 6, 2004). Having a coherent style and image is necessary to demonstrate celebrity status.

Third-season runner-up Diana DeGarmo was an interesting case of an image change. At her audition, she gave the impression of being a young girl, explaining why she loved her pink and black signature colors. The judges debated her image at the audition, with Simon especially disliking her cute, innocent young girl image, a point he reiterated at the end of the season as well. By Hollywood week, she had dropped the pink and black outfits. However, the conversation about her image only appeared in her first episode and her last. She abandoned her preteen style by appearing mature, wearing adult gowns, and in general looking her age or older. She morphed into an elegant woman with a powerful voice, capable of singing ballads and upbeat pop songs. By the end of the season, the bopping teenager was gone.

LIFE TRANSFORMATION

Not only do physical, sartorial, and image changes occur, fashioning the contestants into celebrities, but their lives change as well. The contestants discuss how their lives change from those of normal people into lives of celebrities. This discourse allows the audience to know that the contestants are celebrities—they offer evidence about how their lives have changed. The contestants leave their old lives behind, have fans, and perform celebrity work instead of previous jobs. Kristy Lee Cook noted the affect of *American Idol* on her life: "It's been a huge change for me, because I'm going from such a small life to this huge world that I never would have been introduced unless it wasn't for *American Idol*" (March 12, 2008).

Each season, the show airs packages of the top three contestants visiting their hometowns and being received as celebrities. The discourse of these visits expresses the contestants' life alterations, discursively positioning the contestants as celebrities. Cheering throngs, crying teens, poster-waving moms, and local reporters lobbing easy questions greet them. In the manner of conquering heroes, the contestants drive around town in limousines, ride in parades, and visit former schools. Carrie Underwood's visit home during season four was typical. First, she noted: "Before *American Idol* I had never been on a plane before and now I get to fly home on a private jet. It's awesome." Upon arrival in Oklahoma, she was driven to the local FOX studio to appear on the local news. Fans greeted her at the station with signs, balloons, screaming and applauding enthusiastically. The adoring audience indicated her celebrity. After the interview, she participated in a parade through her hometown of Checotah where fans lined the streets to catch a glimpse of her. "This is crazy, I can't believe it!" Carrie exclaimed. The parade ended at a rally, where Carrie sang the national anthem, and the mayor announced that the governor proclaimed that day Carrie Underwood Day. The mayor also gave her keys to the city and a certificate of achievement from a university, and proclaimed she was going to be inducted into the Oklahoma music hall of fame. Her thoughts? "The parade and autograph signing were absolutely insane. There were so many people there." She then visited her parents' home to see family, friends, and pets. Alas, she couldn't spend the night. "Everyone here is so wonderful. I love this place, but it's time to get back to work" (May 18, 2005). According to the show, the tempo of celebrity life is unceasing.

Season-five contestant Chris Daughtry also spoke about how his life was changed by being on the show:

> It has been one of the biggest life changing experiences I've been through. It's definitely been an adjustment, getting used to the Hollywood lifestyle. You know, I'm used to getting up every morning, spending most of my day working as a service advisor, with very little time living the dream. Now it's totally flip-flopped. My kids are like the most popular kids in school. I'm spending all my time living the dream and the little work I am doing, with rehearsal, I don't really consider it work 'cause I'm doing what I love. It's a huge deal. You know, you're used to living this normal life and now everybody knows who you are. I can get used to it, but it's a little crazy. (March 21, 2006)

American Idol discourse confirms Chris as a celebrity.

Life for the contestants changes as they become celebrities through the show. Never again would Brooke White be a nanny, as she explained, saying,

> Previously to *American Idol* I was a nanny, watching two beautiful baby twin girls. It was a wonderful job and I do miss them. Since *American Idol* life has changed big time. We're doing photo shoots, and we're doing interviews. And everything has changed. (March 11, 2008)

Service industry jobs were out and a lifestyle of fame was in.

David Archuleta continued Brooke's line of thought saying, "It's crazy how much your life changes through all of this. Because one day you're going to school and the next you're doing photo shoots, interviews, and singing" (March 12, 2008). The work of a celebrity was a positive theme. Photo shoots and interviews were never mentioned negatively; instead, celebrity work was intimated as fun and a joy to perform.

David Hernandez reiterated this point:

> My life has changed so much because of the show. I mean, nothing is the same. I started singing when I was six years old, you know, so I feel like it's always been in me, I can't really deny it. It's all I have. Getting up and doing something I love every morning: I couldn't be happier. This is like the point of my life I've been waiting for. I'm ecstatic about it and really grateful. (March 12, 2008)

Celebrity work, fans, and lifestyle aren't the only new aspects of the contestants' lives. The changes occur not just to the exterior aspects of life, but inside the contestants as well. John Stevens explained that "Lately, my confidence has been boosted. I've definitely got more attention from the female population at my school. I guess I'm gradually coming out of my shell" (March 23, 2004). Fame gave confidence to John Stevens, changing who he was as well as his outer appearance.

CELEBRITY DISCOURSE: FAN ADULATION

Another facet of the contestants' celebrity-verification construction is the adoring fans. The screaming fans, the signs they hold, and of course, the numbers of times they vote all contribute to the conception of the contestant as a celebrity, worthy of adulation. The reaction shots of the applauding audience buttress the transformation from a nobody into a celebrity. The more applause or votes the contestant receives, the more talented or worthy he or she seems.

Also, seemingly handmade signs in the audience during the performances add to the conception of fans supporting the contestants' metamorphoses into celebrities. A celebrity cannot be a celebrity without people applauding and approving of him or her.

Season-four contestant Bo Bice described his life transformation as including a new element—fans: "It's really cool, you know, I get all kinds of people that come up to me. It's great. Before *American Idol* the only people that really wanted my signature were the bank and the mortgage company" (May 3, 2005). Fans supplement the creation of celebrity and are a component of life change through adulation and reduction of anonymity.

In every show, there are always sweeping camera shots of the crowd watching the competition. Often a family member or friend of the contestant is singled out for a close-up. The cameras capture how the fans dance, sway, clap, scream, and cry. The most memorable of all the fans may be season six's Sanjaya Malakar fan. Thirteen-year-old Ashley Ferl cried repeatedly whenever Sanjaya sang. Ashley received a lot of airtime and press coverage as Sanjaya's biggest fan. Her apparent devotion to Sanjaya proved he did actually have fans and augmented his claim to celebrity.

Obviously, the first-season contestants were the earliest to realize the importance of the fans. After Simon gave Justin Guarini a negative review, Justin asked the audience, "What did you guys think?" (July 23, 2002). Women of all ages in the audience screamed excitedly in response. Justin recognized the dual significance of the fans: not only in this instance do fans vote for their favorite contestants, but they also build celebrity through adulation. Fan veneration results in greater fame for the celebrity, through magazines sold, television shows watched, and other forms of consumption.

Screaming female fans, while not unique to *American Idol*, certainly play a part in creating a vocal audience, confirming their favorites. Male contestants walking onto the stage often greet a chorus of shrieking female voices from the audience. Clay Aiken, Blake Lewis, and David Cook, among others, had loud female fans in the audience. David Archuleta, particularly, had teenage fans that emitted piercing squeals whenever he took the stage. Their shrill passion was unavoidable, so much so that Ryan commented on it when announcing voting results, saying, "Screaming girls are something you could get used to—and you have to, buddy, 'cause you're going on the tour. You're in the top ten" (March 19, 2008). Screaming fans were positioned as an inevitable result of talent, something that David would have to get used to as his star continued to rise on tour. In this case, however, fans are not only a result of fame, but a cause as well. Those keen fans kept their favorite in the contest, bought his songs on iTunes, and watched the show, combining to build David's celebrity status.

Fan adulation for David Archuleta, positioning him as a celebrity, was extremely apparent on his visit home in the May 14, 2008, episode. Everywhere David went, from the FOX news studio, to the mall, to his old high school, hordes of yelling girls followed and cheered for him. There were seemingly thousands of teenage girls grabbing at him and screeching. He started to cry: "I can't believe so many people came out to this. It makes it all worth it, knowing people appreciate this, all the hard work I've been doing for this. Gosh, I didn't mean to cry," he said through tears, red-faced. It appeared David recognized the importance of his fans; they were the reason he attained celebrity and he was grateful for them. The fans were so vehement in their love and support of David that he became emotional too.

Aspects of fan worship, particularly the homemade-looking signs in the theater audience, impressed other contestants as well. Every season, signs proliferate throughout the theater audience, popping up for every contestant. Season-five contestant Kevin Covais said, "It's amazing, this whole experience has just been incredible but one of the best things is having fans out there with signs. It's really cool" (March 14, 2006). The fans contribute a personal element to being famous. A fan (if he or she did make the sign) taking the time to create a sign demonstrates that he or she feels a connection to the contestant.

Signs illustrate the affinity fans hold for contestants, as well as bolster celebrity through demonstrating fandom. Signs from season seven included "Hook'd on Cook" and "Cougars for Cook." In one season-six example, Ryan Seacrest pointed out a sign in the theatre audience that said "Mandisa will you marry me?" Ryan said to the camera, "Mandiva has arrived" (March 21, 2006). Perhaps people watching at home may have missed seeing the sign, but Ryan ensured they were aware that Mandisa was a celebrity with fans who cared enough to make signs for her. Moreover, calling Mandisa a diva, as the name Mandiva implied, granted her the connotations with the idea of a diva.

The fans build celebrity through the public admiration illustrated through signs, and also through their sheer numbers. There are hundreds of thousands of fans of *American Idol* and the contestants. On the May 20, 2003, season-two finale, the show alternated between showing the theatre in Los Angeles and fan gatherings in Raleigh, North Carolina, Clay Aiken's hometown, and Birmingham, Alabama, Ruben Studdard's hometown. Almost nine thousand screaming Clay fans were in the Raleigh arena, including several of Clay's self-anointed closest friends. As Ryan said, it was "Clay mania" (May 20, 2003). The fan turnout and their obvious enthusiasm meant that even if Clay didn't win, he was still a celebrity.

Fans also contribute to contestant success by voting. On season three, third-place finisher Jasmine Trias had immense audience support. The size and fervency of her fan base astounded her. During her visit to her home state of Hawaii, she stated, "I just came from the airport, it was just so crazy. It was so insane. I can't believe how much support I had here. I mean, I know how much support but it just totally hit me today." She met Hawaii's Lieutenant Governor, "but nothing could prepare me for the overwhelming response at my high school. Then it got even crazier when over eight thousand people showed up to see me off" (May 18, 2004). After being voted off the next night, she acknowledged her fan base:

> I just want to thank all my fans for believing in me and for embracing my talent and for making my dreams come true. This, top three, I mean I could not ask for more. You know. It's been such an honor to share the aloha spirit with the rest of America. And I'm just so glad that I had this opportunity. (May 19, 2004)

The enthusiastic devotees essentially made her a celebrity. Without fans, a person cannot be famous. Arguably, supporters are the most fundamental component to fame.

Many of the contestants realized that they owed their celebrity success to the fans and thanked them. As Kristy Lee Cook said, "Thank you, America, for voting for me. This is a huge dream and it never would have come true if it wasn't for all the fans" (April 15, 2008). Kristy acknowledged her fans for making her celebrity dream a reality. Similarly, fifth-season contestant Ace Young, upon being voted off the show, said, "I've had a blast. I can't wait to perform with the top ten and see all my fans that have seen me through the TV" (April 19, 2006). On tour, Ace could truly enjoy his celebrity standing in a tangible format, with thousands of screaming fans before him, in opposition to the mediated version of celebrity. The fans create celebrity through the support and consumption of contestants and products. Without fans, celebrities would cease to exist.

Thus, *American Idol* uses authenticity portrayals and celebrity discourse to engender contestant celebrity. The contestants are both ordinary through authenticity and extraordinary through the celebrity discourse. This demonstrates a tension between the intimate, attainable celebrity and the high-aura, unreachable distant celebrity. The *American Idol* celebrity is both instantly recognizable as average and exceptional through talent and celebrity lifestyle. This is a new type of celebrity, amalgamating the manufactured reality-TV renown and talent-merited fame. Simultaneously, *American Idol* celebrity is manufactured through discourse and warranted through talent.

Chapter 7

Show Discourse and Collective Identity

Within the first seven seasons of the program *American Idol*, discourse engendered various collective identities. Collective identities, the shared identities of the fan communities, form through show discourse and fan identification, categorization, and comparison. This chapter specifically uses social identity theory to understand how discourse helps generate collective identities associated with the program. According to Morley and Robins (1995, p. 72), "Collective identity involves the achievement, by individual actors or by social groups, of a certain coherence, cohesion, and continuity." Morley and Robins (1995, p. 72, emphasis in original) also claimed that "collective identity must be sustained *through time* . . . it must also be maintained *across space.*" Likewise, Anderson (1983/2003) asserted that some collective identities are "imagined communities," imagined because people can never meet most other members, yet the identity remains communal in concept. These collective identities are often created and maintained through media.

Undeniably, an essential aspect of the creation of national or cultural identity is some form of "homogeneity—ethnic, religious, linguistic, cultural, territorial" (Morley & Robins, 1995, p. 188). There must be a linkage between people to form a collective identity. As an example, America contains a plurality of ethnicities, religions, and languages. However, everyone in the lower forty-eight states is linked through geographic placement. Physical place functions as a connector, establishing a common element despite other differences, uniting people in America into an imagined community (imagined because people will never meet all of the people within it).

Nonetheless, physical space need not be the primary connector. Collective identity can transcend boundaries through media, which are not stymied by geographic boundaries. For example, expatriates reading their hometown newspaper are connected to their nation and belong to a nationality through the newspaper, its language, and the events, emotions, and values within it. Even before the invention of Gutenberg's printing press, media have been transportable, capable of traversing distance and time, spreading knowledge. Correspondingly, Anderson (1983/2003) observed the spread of national or ethnic identities throughout areas with varied political ideologies. Global mass media transcend borders and help inculcate belonging, identification, nationalism, and community (Price, 1995).

Social groups continually evolve and adapt, yet maintain a foundation (Schlesinger, 1991). This foundation of social groups is the basis of a collective identity: "Identity is a question of memory, and memories of 'home' in particular" (Morley and Robins, 1995, p. 91). A single conception of "home" may be identifiable and the basis of the collective identity. Home could be a physical place, spiritual beliefs, or essential cultural elements—all may be indicated not only through cultural transmission of education, home life, or traditions, but also through mass media (Price, 1995). Home may be considered anything that links a community together, fostering a collective identity. For some people, *American Idol* may function as a home, linking people together.

SOCIAL IDENTITY THEORY

This study uses social identity theory, formulated by psychologist Henri Tajfel, to describe the formation of groups associated with *American Idol*. Once people belong to a communal group centered upon a shared interest, they share a collective identity.

One basic premise of social identity theory is that humans "are social beings who derive an important part of our identity from the human groups and social categories we belong to" (Tajfel, Jaspars, and Fraser, 1984, p. 5). Almost any variable can "function cognitively as a criterion of social categorization to produce an awareness of shared social identity" (Turner and Bourhis, 1996, p. 34). For instance, shared goals, ethnicity, or perceived threat could all be causes for shared social identity. Additionally, interdependence is common within groups (Turner and Bourhis, 1996).

There are three primary facets of social identity theory: categorization, identification, and comparison (Turner, 1996). First, categorization is decided by the attributes an individual accords a social group, such as ability to discern talent or congeniality or any other quality. Next, identification occurs when a person decides to join and then identify herself through the group. Lastly, comparison reveals the positive differences the person discerns between her group and others. The motivation for the process is to attain a positive identity (Hogg, 1996). People desire to belong to a group that they consider to be positive, in order to bolster self-esteem (Abrams, 1996).

Abrams (1996, p. 144) claimed that social identity theory is "an engine for collective attitudes and behavior." Group members tend to share beliefs and act similarly. Moreover, collective identities provide a shared social construction of reality for members (Abrams, 1996).

Social Identity Theory and Collective Identity

Many authors have noted that identity is constituted through differentiation, noting an us-versus-them comparison related to collective identity. For example, Morley and Robins (1995) alleged that Europeans understand their culture through comparison to American culture. The American dominance of international media markets created an anti-American emotion and therefore reinforced European collective identity. Comparison through media solidifies communal identity (Morley and Robins, 1995) and provides a social organization structure (Schlesinger, 1991).

Audience Collective Identity

Audiences are generally composed of people with a shared interest in a media text or celebrity. Because of this communal identifying trait, audience members share at least one specific collective identity—for example, as an *American Idol* fan. Gamson (1994, p. 132) noted, "Watchers are connecting with each other through the 'sport' of sighting, identifying, and categorizing celebrities, exchanging bits of information, through their common experience of and role in the spectacle." While he was discussing audiences located in a physical place together, this could also be applied to virtual spaces.

Reality TV: Collective Identity

Media consumption also builds a community through the immediacy of viewing. There exists an "imagined community of viewers *at the moment of watching*" (Kavka and West, 2004, p. 140, emphasis in original). Similarly, the live nature of the media event "draws the audience into the momentary constellation of a television community" (Kavka and West, 2004, p. 140).

Kavka and West (2004) cited the example of Princess Diana's funeral as a community-generating experience. While this event was highly emotionally charged, it is possible to generalize from it. The veracity and reality of an event draw a tighter bond among viewers, no matter the level of emotions involved. Combined with the actuality of an event, the ahistorical and atemporal nature of reality TV contributes to the formation of community (Kavka and West, 2004).

Cavender (2004) noted that *America's Most Wanted* and *Survivor* both create a televisual community of audience members through the action of watching the programs, as well as through the narrative devices within the shows. Within *America's Most Wanted*, the host speaks directly to the camera, and thus the audience, facilitating a communal emotion in viewers. Similarly, the host of *Survivor* frequently uses language that invokes community (Cavender, 2004). However, Cavender asserted that the depictions on the programs actively degenerate traditional communities, instilling fear and cynicism within American culture.

In Cavender's (2004) argument that reality TV serves to "erode and undermine" community, he cited works by Riesman (1950), Putnam (2000), and Etzioni (1993) that identified the waning of American civic commitment, lack of social participation in local communities, and increases in crime as indications of traditional communities' decline.

Cavender claimed that *Survivor* featured a dystopic society, while *America's Most Wanted* featured suspicion—everyone was a potential criminal or victim. On *Survivor*, the portrayed community is filled with lying, whispers, and secrets. While both shows espoused a traditional concept of community, as "a coherent unity," their depictions presented a negative alternate version of community, filled with "larger social anxieties" that translated off screen (Cavender, 2004, p. 170). Nevertheless, while Cavender sees a detrimental effect on geographic local communities, the rise of reality TV has indeed facilitated new communities' development, not necessarily based on geography. While Cavender's argument is quite persuasive, *American Idol* generates an online community that most likely does not contribute to the detriment of overall American society.

The community fostered by *American Idol* is positive. While the show is a competition, a family metaphor is stressed repeatedly. The contestants every season claim to truly like each other and the judges. They often smile and giggle at each other. Family members are frequently shown claiming how proud they are of their relative on the show. Some common emotional attributes expressed on the show are joy, pride, happiness, and friendliness. Following Cavender's argument, the offscreen communities should have reflected those qualities. The following chapter discussing the online message boards finds this to be true.

Like *Survivor* and *America's Most Wanted, American Idol* also featured direct-viewer address and narrative devices creating an audience community. In addition, the interactivity between the viewer and the program of *American*

Idol fostered collective identities of groups. Viewers choose to participate in the media of celebrity because "the draw is not only to witness, but to help create the spectacle" (Gamson, 1994, p. 137). Helping to create the spectacle, belonging to the group that expands to include not only audience members but also those working on and for the show, cultivates a particular form of collective identity. Voting and discussion boards permit audiences to identify as fans and as a part of the collective identity of *American Idol*. Participation, as well as liveness and discursive functions, create a unique *American Idol* collective identity.

As Foster (2004, p. 279) stated, "A show's fundamental meaning must dovetail with the dominant meanings of its audience for it to be compatible with the lives of its viewers." Foster (2004) contended that *Survivor* is a microcosm of American culture, particularly the Horatio Alger narrative of self-reliance and hard work ensuring success. Perseverance and luck can "propel anyone onward and upward and provide hope and promise in the face of hardships" (p. 280). On *Survivor*, lying and strategy replace honesty and hard work. Western values of toughness, alliances, and strategic friendships bring success (Foster, 2004).

While many factors affect the success of a program, including audience demographics and scheduling placement, Foster (2004) noted that *Big Brother* never achieved a comparable level of success in America that it achieved in the United Kingdom. This, along with the success of *Survivor* in America, may be related to the cultural morals and values represented in the shows. Audience expectations and taste in television have been shaped through years of programming—they may differ according to culture. Following this logic, *American Idol's* rating success and popularity may be related to its illustration of cultural values and ideals. Each particular franchise of *Idol* is tailored to the specific location where it airs. For instance, *Ídolos (Brazil)* may reflect the Brazilian culture, just as *American Idol* reflects an idealized American culture.

COLLECTIVE IDENTITY AND *AMERICAN IDOL*

Collective identity as related to *American Idol* was apparent through the show and message board discourse. While the practice of watching the show live may be a ritual that helps form collective identities, *American Idol* also constructed certain collective identities through discourse, both verbal and visual. The show assumed that the collective identity of a portion of the audience exists; in a cyclic fashion, because the show speaks about the identity, it did exist—at least on the show. Discourse about a collective identity ensures and maintains its existence. The following chapter concerning fan collective identities visible on the message boards reinforces this concept.

The discourse analysis of the first seven seasons of *American Idol* revealed three primary show-constructed collective identities: regional, contestant-based, and American.

Regional Collective Identity

Verbally and visually, the show hosts, judges, and contestants often referred to contestants' hometowns and areas of origin, facilitating identification with a region for both the contestant and fans. Corresponding statements and expressions on the show formed and maintained assorted regional collective identities, with one in particular based upon the American south. This created a way for audience members to remember each contestant, differentiating the contestants from each other. Moreover, fans may have identified with the same region as a contestant did. For example, people who live in the southern United States may have voted for Kelly Clarkson because she is from Texas. Furthermore, people may consistently identify over the years with different contestants from their general area.

Regional Discourse

The discourse on the show repeatedly associated contestants with their specific areas. For instance, on the first-season finale, Ryan said, "In a few minutes, either the boy from Philly or the girl from Dallas will be named the American idol" (September 4, 2002). Discourse like this occurred throughout every season. Additionally, contestants identified themselves through their hometowns as well.

On season three, Latoya London asserted, "I'm from Oakland, California, and I've been a Bay area girl my whole life" (March 16, 2004). Season-five competitor Mandisa said, "I am trying to represent for Nashville. I'm trying to hold it down for y'all" (February 28, 2006). Mandisa was actually speaking to the people in Nashville, telling them she was attempting to compete for them, encouraging a fan group to form. Similarly, Vonzell Solomon reminded the audience she was from Fort Meyers, Florida, while Trenyce informed viewers she was from Memphis.

Some contestants took this concept further, identifying with a small-town, rural, or country lifestyle in general. Audience members, if not identifying with the regional area, may have identified with the concept of a small, wholesome town. In season seven, Amanda Overmeyer said, "I'm from Mulberry, Indiana. It's a small rural town outside of Lafayette . . . Yeah, it's pretty boring where I'm at, just country rural, you know" (March 11, 2008). Kristy Lee Cook was also from a small town. She elaborated:

> I'm from Selma, Oregon. It's a really small town, there's maybe like five hundred people. It's like if you blink your eyes, you're already through the town. This has been such a change for me. It's totally opposite what I'm used to. I mean, I'm a country girl and to come into the city and be in the top twelve, it's a news flash. (March 11, 2008)

The South

Many contestants were from the American south and emphasized this throughout the seasons. For example, in an example of discourse not spoken by the contestant, Randy identified Bucky Covington as southern. He commented after a performance, "Bucky, Bucky, like me, man, representing the dirty south" (February 28, 2005). Bucky had identified himself as southern previously, and Randy supported this label. There are a plethora of examples analogous to this one.

Furthermore, not only did some contestants identify themselves as southern, but they also overtly identified with southern culture. A case in point occurred at Diana DeGarmo's audition that aired January 28, 2004. Simon said to Diana, "The smiling's quite irritating." She replied, "I'm sorry, it's a southern thing." Later in the season, Diana happily discussed her hometown, saying:

> I've lived in Georgia about thirteen years now, so about almost all of my life and I love it here. The town I live in is called Snellville. It's great! I think we have a cool motto: 'Where everybody's somebody.' See, I'm trying to prove that motto right, that's the whole reason I'm doing this. (March 16, 2004)

Fourth-season competitor Lindsey Cardinale also claimed to be from a small southern town, saying in a statement similar to Diana's, "I'm from Pontucla, Louisiana. And it's a small town, but we have a huge strawberry festival every year. We're known for our strawberries. But I wanna put 'em on the map for something other than that—me" (March 1, 2005). Both Diana and Lindsay used the inclusive word "we" when speaking about their hometowns. They identified with other people living there or from there, as well as other people from small towns.

Ruben Studdard

Ruben Studdard was an exceptional example of facilitating a collective identity based on a geographic area. The second-season contestant fostered an avid-fan collective identity. Fans of Ruben as a contestant and fans of his home region, Alabama, coexisted to create a collective identity based upon both Ruben and Alabama. Throughout season two, Ruben appeared again and again wearing jerseys and shirts with the 205 area code for Birmingham, Alabama, on them. Additionally, he repeatedly and proudly asserted his love for Alabama. He even managed to sing "Sweet Home Alabama." As Ruben promoted his affiliation with his hometown, fans mimicked him, appearing wearing similar shirts with the 205 numerals, which also appeared on hand-held signs.

On the February 11, 2003, show featuring the top thirty-two contestants, Ruben explained his jersey:

> The shirt that I'm going to be wearing on the show is just a shirt that I got made to represent where I was from and just to represent my flavor. So, I got 205, which is my state's area code, and I got flavor on the shirt. This is my dream and I hope you like it, America.

By April 1, 2003, people in the theater audience were wearing 205 shirts, signifying their collective identity membership.

Teddy bears had a function similar to the jerseys. Gladys Knight had bestowed the moniker of the "velvet teddy bear" upon Ruben, and his fans embraced the metaphor. Audience members in the theatre held up teddy bears to show their support of Ruben and their membership in the Ruben-affiliated collective identity.

The season-two finale aired May 20, 2003, and featured two dynamic performers, each with a large and vocal fan base: Ruben Studdard and Clay Aiken. While Ruben had what Ryan Seacrest called the "205 nation" backing him, Clay had a rabid group of fans called the Claymates. According to Ryan, "We've split the nation right down the middle, between Ruben fans and Clay fans." This statement implied that the entire nation watched *American Idol*. Everyone was united into one audience of the show, a concept that is discussed later in this chapter. For the season-two finale, the two fan collective-identity groups battled.

On the finale, Clay's and Ruben's hometowns were shown. Clay's fans gathered in the RBC Center, an arena in Raleigh, North Carolina, while Ruben's fans gathered in a church in Birmingham, Alabama. Ryan sat in each finalist's dressing room and showed each his hometown through a live video feed. He introduced Ruben's fans by saying, "Let's go to two oh five, Birmingham, Alabama, for one second." Inside the church, Ruben's fans cheered and screamed. Several were wearing 205 jerseys. Similarly, in the RBC center, thousands of fans held signs for Clay while cheering.

In 2007, on *American Idol Rewind*, Clay narrated the events of the evening, explaining, "As the pressure-packed performances came to a close, the fate of the contestants were now at the mercy of their passionate fans. It was the Claymates versus the 205 for the title of American idol" (April 26, 2007). This identified the two fan groups: the Claymates, sharing a collective identity based upon Clay, and the 205, based upon Ruben and his hometown. In the Kodak theatre, the audience appeared split, with several signs for each finalist and many people wearing 205 jerseys. While Ruben won in a close race, each contestant had solidified the collective identity of his fan group.

Jasmine Trias

Like Ruben Studdard, another contestant fomented a remarkable collective identity: Jasmine Trias. Jasmine actively promoted Hawaiian culture. Jasmine took Ruben's hometown affiliation concept and enhanced it by continually putting forward Hawaiian cultural concepts.

Through her statements, Jasmine strongly associated herself with Hawaiian culture. Early in season three, on March 2, 2004, she explained, "Aloha! Woo! My mission is to bring Hawaii to the rest of the world. And I want to teach everyone how to do the hula." As Jasmine spoke, images were shown of her giving the Hawaiian hang-ten symbol and of her teaching the hula to other contestants, a vocal coach, and stylists. A couple weeks later she elaborated on her connection to Hawaiian culture. Jasmine said:

> My grandma enrolled me in hula classes at the Hawaii plantation village in Wa Pahu and I've been there ever since. Hula dancing is like telling a story with your hands. It connects you to the history of the Hawaiian Islands and to their culture. So I think it a very important thing to know, especially if you're a Hawaii resident. The reason why I wear the flower in my ear—it's my Hawaii thing and if you all want to know, it's right ear single, left ear taken. Through the first audition, I had the flower in my left ear. And now single! (March 16, 2004)

The flower she always wore behind her ear was a telling example of her symbolic link to Hawaii. One week, she neglected to have it in her hair and Simon commented, "Well, Jasmine, I'm glad we've lost the silly flower at last." She rejoined, "No, it's still here. Right here (patting flower on her hip)" (April 20, 2004). The flower represented Hawaii and its culture. As her mother said, "She just loves the Hawaiian culture" (March 23, 2004).

Not only was Jasmine connected to Hawaii, but the show discourse indicated her fans were as well. In an unexpected twist, Latoya London was eliminated and Jasmine entered the top-three competition with Diana DeGarmo and Fantasia. Upon Latoya's elimination, Simon said, "It's a surprise. I think,

Jasmine, you have a lot of thank-you letters to write to Hawaii." Simon elaborated the following week by saying:

> Jasmine, you're a lovely girl. But the reality is, in my opinion, the fact that you're so young and so sweet and you have this unbelievable support from Hawaii is why you are still in the competition. For me, it has to end tonight. Sorry. (May 18, 2004)

Upon her elimination, Jasmine referenced her connection to her fans and Hawaiian culture:

> I just want to thank all my fans for believing in me and for embracing my talent and for making my dreams come true. This, top three, I mean I could not ask for more. You know. It's been such an honor to share the aloha spirit with the rest of America. And I'm just so glad that I had this opportunity. (May 19, 2004)

Contestant Collective Identities

Not only were collective identity groups regional, they were based around contestants themselves. Season-two runner-up, Clay Aiken, had a large self-identified group: the Claymates. The Claymates lacked a regional affiliation; rather, they simply identified as fans of Clay. They were spread throughout the country, with their collective identity transcending physical space. Ruben's fans may also have lived throughout the country, but were associated with Alabama through discourse. The Claymates were repeatedly referred to during the show discourse and self-identified in the audience through signs.

For example, in season three, Clay returned and performed a song from his new album. Ryan noted, "the Claymates are here tonight" to screaming and applause (March 17, 2004). People in the audience grasped signs and appeared wildly enthusiastic—and this wasn't even Clay's season.

During 2003, when Clay appeared on the show, it was clear that he had a very passionate and fervent fan base. Whenever Clay sang, audience members waved signs and appeared moved by his performances. Indeed, years after Clay's *American Idol* season, there still exist quite a large number of fan sites and message boards dedicated to him. His fans frequently label themselves as such.

Toward the end of season two, on May 19, 2003, with Clay onstage next to him, Ryan pulled a Clay fan out of the audience named Erin. She proceeded to answer arcane trivia questions about Clay's life. While Erin may have been prompted earlier on the questions, Clay appeared totally surprised that someone knew so much about him. Erin knew how much a red leather jacket that Clay had worn in an earlier performance had cost, his old nickname at the YMCA, and how old he was when he performed as his high school mascot. Clay appeared stunned and baffled by the extent of her knowledge. Erin was the epitome of the Claymates—obsessed with Clay and eager to learn the details of his life.

Throughout the years, discourse on the show aided in the formation and sustenance of contestant-based collective identities. Fan identification manifested in different ways. For example, season-three contestant Amy Adams fans (who appeared to be mostly her family) demonstrated their membership in a fan-group collective identity by dying a pink streak in their hair just like hers. While the audience members themselves added to the discourse visually, Ryan often verbally pointed out their existence.

Ryan's confirmation of the fan collective identities helped create, popularize, and sustain them. In this way the show strategically used identity, providing labels for the audience to potentially embrace. In season five, Ryan interviewed Elliot Yamin onstage and noted his fan base: "The Yaminians are cheering 'cause Elliot is sitting with me on the stool" (April 25, 2006). Later in the season, he asked Chris Daughtry, "Each of the contestants have fan clubs. What are your fan clubs named?"

Chris replied, "Oh man, we got the Daughtry gang, Chris's crew, the chrisaholics, the list goes on" (May 9, 2006).

While many contestant-based collective identities appeared to occur with minimal aggrandizement by the contestants, another noticeable example of a strong collective identity generated and reinforced through discourse occurred in season five: Taylor Hicks's Soul Patrol.

While Clay seemingly did little to encourage the collective identity of the Claymates, Taylor actively helped create the collective identity based upon him as a contestant, the Soul Patrol, through his consistent statements and referrals to it. A collective identity based upon him could only help Taylor in the competition. Throughout season five, after performing and as Ryan informed the audience which number to call in order to vote for Taylor, Taylor would shout "soul patrol, soul patrol, soul patrol." Whenever he could, Taylor referred to the Soul Patrol. For instance, answering a question, he once pointed to the audience and into the camera while saying, "I chose 'Try a Little Tenderness' by Otis Redding. It's got a lot of heart and soul and I'm doing it for the Soul Patrol" (May 16, 2005).

While Taylor promoted the Soul Patrol, the show did so as well by selecting the footage that aired. On the top-three-finalists episode featuring Taylor's visit home to Birmingham, Alabama (the same small hometown shared by Ruben Studdard), Taylor was shown consistently yelling "Soul Patrol." As he got into his car, he said, "The Soul Patrol rides on." During the parade through town he stood through the car sunroof and shouted "Soul Patrol" to bystanders. They shouted back to him, "Soul Patrol!" Upon meeting the governor of Alabama and his wife, Taylor said, "I've got two new inductees to the Soul Patrol right here" (May 17, 2006). The constant support and substantiation of the Soul Patrol's existence helped to maintain the high level of support for Taylor. Similarly, his highly visible validation of the Soul Patrol collective identity, as well as his occasional inclusion of himself within it, added to the high excitement level and most likely the number of members.

OVERALL AMERICAN COLLECTIVE IDENTITY

While off-screen American identity is complex and multi-layered, the show discourse formulated an ideally positive American identity, free of complications. The show identified the audience as American through discursive devices that included consistently addressing the audience as American, pronoun use, referencing American themes, and visual cues. The visual cues corresponded with Billig's (1995) conception of banal nationalism: everyday reinforcement of national identity through material culture and references that people are accustomed to and no longer notice, such as American flags hanging in post offices. This occurred in the opening sequence of every show: the use of American flags, and the primary colors of red, white, and blue. The discourse also referred to other American themes, which contributed to the overall American identity of the show and the audience. Knowledge and understanding of American symbols, concepts, and ideals depicted on the show all facilitated the construction of an American collective identity. Visually, the show discourse referenced many American concepts, including places and significant American material culture. Through this and other discursive tactics, *American Idol* constructed an inclusive American collective identity for the viewing public.

The most important discursive feature positioning the audience into a single cohesive group, and then a subsequently single American identity, was the breaking of the imaginary fourth wall. Throughout the first seven seasons, everyone spoke directly to the audience. Ryan frequently said, "Welcome to *American Idol*" directly to the camera and hence people at home, creating a cohesive audience by addressing the particular group of people watching the show. In another example, Ryan spoke straightforwardly to the at-home audience through the camera, saying, "I hope you're ready for the challenge, because the judges are relinquishing control and handing things over to you. Your first task is to get us down from the remaining twenty-four people to our top twelve" (February 19, 2008). Ryan's use of pronouns and speaking directly to the audience placed all audience members within the same group. He called on the audience as a cohesive unit, not as individuals.

Furthermore, throughout the first seven seasons, the host, the judges, and the contestants repeatedly labeled the audience watching as American, as did the very title of the show. While seemingly an obvious aspect of the show, the means by which this construction occurred merits attention. The hosts and judges often labeled the general audience as American through verbal utterances such as "Welcome back, America" and "America, you decide." The judges said phrases like "I'd like to congratulate America once again for getting it absolutely right." Occasionally Ryan or the judges expressed that they were American, too (except, notably, Simon). For example, Paula once said, "We're all responsible for putting these wonderfully talented contestants through. It is about connecting with America, the TV viewing audience" (February 28, 2008). Paula simultaneously asserted that the TV audience was American and placed herself in the American collective identity by using the term *we*. Particularly during season seven, Ryan often stood in the crowd, speaking to the camera and the at-home audience while surrounded by studio audience members. This connoted that Ryan was one of the audience and part of the collective identity. On the first season, cohosts Brian Dunkleman and Ryan Seacrest said phases such as "America voted . . . and agreed" or "America, you choose who continues." By the first season finale, the audience held signs that incorporated American flags, and red, white, and blue colors. And upon Kelly's win, Ryan said, "Congratulations, America. You've made this happen, you have found your idol" (September 4, 2002). He congratulated the audience, not Kelly.

In season two, Ryan reminded the audience at home who they were by introducing those at home to those in the studio: "America, this is the studio audience. And studio audience, please say hello to America" (April 15, 2003). Similarly, the season-two finale had the theme of the entire country being united as *American Idol* fans. Throughout the entire show, Ryan declared each state that Ruben or Clay won, positioning the contest as a presidential election:

> Good news for Clay. You have taken the state of North Carolina, winning those votes there in North Carolina, so congratulations. One down, still a few more to go. It is like a presidential election. I think we actually have more votes for this. (May 20, 2003)

American Themes

Each season, the auditions occurred in several different American cities, each with its own symbolic contribution to American culture. The St. Louis episode incorporated the famous arch, the Dallas segment featured cowboys, and the Philadelphia audition highlighted the Liberty Bell. Verbal and visual discourse also emphasized the American nature of the audition cities. In every city, American flags flapped in the breeze.

Moreover, some contestant song choices were overtly American and patriotic, espousing unconcealed American ideals. On season seven, Kristy Lee Cook sang, "God Bless the USA" under red, white, and blue lighting. Also on season seven, David Archuleta sang "America" by Neil Diamond, another patriotic song about immigration and the American dream. While each sang, a huge digital American flag waved on the giant television screen behind him and her.

Another American aspect of the program is the English language. Rarely was any language spoken except English, despite contestants having other language skills. Although three contestants were of Filipino ancestry, Tagalog was never spoken on the show. Even songs with lyrics in other languages, such as Spanish or French, were rarely performed. Similarly, reminiscent of the us-versus-them function of collective identity theory, Simon's British accent was often ridiculed on the show. Southern contestant Kellie Pickler once said to Ryan about Simon, "It's his accent. I don't understand what he says." Simon replied, "Likewise." Kellie retorted, "I don't have an accent" (April 4, 2006). Kellie, of course, had a thick American southern accent. She identified as American (us) and identified Simon as the other (them). Season-seven contestant Michael Johns, originally from Australia, feared such judgment. He once said, "Hopefully, people can judge me on my voice and my passion and not whether I can say g'day" (February 19, 2008).

Outside American Events

American Idol occurred in real time and occasionally referenced events that happened outside of the show, which functioned as unifying events for the collective American audience.

During the sixth season, the Virginia Tech campus murders occurred. On the show that day, Ryan said, "At this difficult time, we want to say to all those affected by the tragedy at Virginia Tech our thoughts and prayers are with you" (April 17, 2007). Ryan's use of the term *we* signified that those working on the show, as well as those watching, were part of the national American community. Additionally, acknowledging that something horrific happened places *American Idol* contextually in the moment, securing *American Idol's* place in the American cultural fabric. Ignoring the event would have placed *American Idol* in the realm of pre-taped sitcoms, unable to address current events. The real-time nature of the show further connected it to the American audience. Just as other Americans felt compassion for the victims, so did *American Idol.*

Happier events were also noted. During the 2008 World Series, commercials aired for the upcoming eighth season of *American Idol.* The commercials juxtaposed both competitions, comparing World Series and *American Idol* highlights. Each commercial closed with Simon saying "America's other favorite pastime" (October 26, 2008).

Idol Gives Back

During seasons six and seven, *American Idol* aired a special two-hour show titled *Idol Gives Back.* This show raised money for various charities, featured numerous celebrities, and called attention to poor people suffering in America and Africa. It also fomented the American collective identity of the audience in several different ways, including linking the audience and the show itself in a common cause.

Ryan explained the special episode thusly:

> Here at the Kodak, three thousand people, thirty million watching at home, and countless *A*-list stars wait in the wings all with one goal—to raise money for disadvantaged children around the world and here at home. Together, we're going to save some lives. (April 24, 2007)

As Ryan noted, all of these people were physically located in America, and connected through space (location) and time (the live nature of the show). Later, Ryan reiterated, "Tonight, the country comes together to see what it can do to change the lives of thousands of people at home and in Africa" (April 24, 2007).

Ryan's use of the pronoun "we" was repeated throughout the show as well. In a pertinent example, Brad Pitt spoke about rebuilding New Orleans. He said, "I do find this about America. When we understand a situation, we rally. We called upon the people of the USA to lend a hand and that's what happened. We're getting it done" (April 9, 2008). This is just a single example of how the *Idol Gives Back* discourse facilitated the existence of a single inclusive cohesive American identity.

Other aspects of *Idol Gives Back* indicated the American identity of the audience. Various American charities were profiled, including Save the Children, The Boys and Girls Club, and America's Second Harvest. These charities help poverty-stricken children throughout America, including in the rural Midwest, Appalachia, hurricane-destroyed New Orleans, and poor areas just outside of Los Angeles. Each of these areas and their inhabitants were profiled on *Idol Gives Back.* These profiles featured a diverse array of children discussing their similar lives, hopes, and dreams. Just as these children lived everywhere in the United States, so did the *American Idol* audience. Through caring for these needy children who live throughout the U.S., the entire audience and participants on the show were integrated into an American group. At one point in the season-seven episode, the contestants even sang a song titled "My American Prayer" about ending poverty. The diverse reality of the poverty locations also added to the American identity—presumably most Americans could recognize at least one of those types of places as somewhere that resonated in his or her life or family history. The show made an effort to enhance the group American identity through personal relevance.

In another timely feature related to *Idol Gives Back*, during season seven, the three American presidential candidates appeared on the show the night after *Idol Gives Back* aired. Hillary Clinton, Barack Obama, and John McCain each spoke to the American audience in taped segments, soliciting money for poor children. Like the presidential race, *American Idol* is a contest in which the winner succeeds by receiving votes from the American public. The appearance of the candidates informed the audience of its American identity and voting responsibility in both contests. Similarly, later in the season, the then-current president and his wife—George W. and Laura Bush—appeared in a highly symbolic taped segment. President Bush said, "We thank all the *American Idol* viewers who have shown the good heart of America" (May 1, 2007).

Celebrities appearing on *Idol Gives Back* ranged from Reese Witherspoon and Brad Pitt to Eli and Payton Manning, representing different aspects of American life, including Hollywood, music, government, and professional sports. There was at least one celebrity everyone in the audience could recognize and admire. Most of the celebrities were Americans.

Additionally, on *Idol Gives Back* the only other countries shown were African. Since only underprivileged and hungry children were shown, Africa was portrayed as a place to pity. America, while not without its problems, appeared superior in feeding poor children and providing education. According to the us-versus-them duality within social identity theory, the show discourse favorably compared America to Africa, further enhancing the cohesive collective identity.

America vs. England

Parallel to the America/Africa comparison, another contrast throughout the first seven seasons was the distinction between America and the United Kingdom, as exemplified through judge Simon Cowell. Following social identity theory, this comparison theme illustrated the positive characteristics associated with the American collective identity.

For example, at the January 15, 2008, Philadelphia audition, the judges heard cheering outside of the judging room when Angela Martin exited holding a yellow ticket. Simon said, "You know what's amazing about this country? It's that you're generally happy when someone you know does well."

Randy questioned, "You're not happy?"

Simon replied, "The idea that someone I know is happy and I celebrate with them? Wouldn't happen." In this Simon-facilitated comparison, American benevolence and empathy clearly trumped British indifference and resentment.

On another occasion, in an exchange that occurred numerous times over multiple seasons, when asked which country has the best singing talent, Simon replied, "I will concede Americans have more singing talent, but on the judging side, the Brits" (March 12, 2008). His humor may have been entertaining, but his answer aided in building a positive American collective identity. Whether Simon was truthful or not, Americans could feel good about being superior to their British counterparts, especially as a British expert himself confirmed that Americans were more skilled.

CONCLUSION

Through the lens of social identity theory this chapter illustrates how *American Idol* discourse fosters collective identities, including an American national collective identity. Additionally, through *Idol Gives Back*, American collective identity was constructed as benevolent, sympathetic, and magnanimous. On the show, Americans compassionately helped poor children, not just in the U.S., but in other countries as well.

Not only was the collective identity American, but the group also represented all of America. The audience members weren't called Americans. They were called, literally, America. Thus, those people within the group who ascribed to the collective identity were actually considered to compose America.

American themes and ideals in the show correspond with dominant American cultural meanings. This correspondence helps build the American collective identity. Equivalent to Foster's contention that *Survivor* is a microcosm of America (Foster, 2004), *American Idol* also represents and constructs superlative aspects of American collective identity and thus culture. Moreover, the accessibility of the dominant American ideals may help facilitate the show's high level of popularity (Foster, 2004). Through *American Idol* discourse, constructed collective identities, particularly an American identity, are formed and maintained. These identities correspond with American ideals of leadership and charity.

Additionally, this chapter agrees with Cavender (2004) that television shows may encourage communities through discourse cues. However, while Cavender (2004) found that *Survivor* and *America's Most Wanted* promote dystopic communities, *American Idol* featured a positive and idealistic view of community. The characteristics displayed on the show were only positive and never negative, despite the fact that the entire season is a competition with one winner. In contrast to *Survivor*, this winner is audience selected; the contestants do not eliminate each other. *American Idol*, unlike *Survivor*, hardly depicted any striving, manipulation, ruthlessness, or aggression. Instead, the show championed the virtues of hard work, practice, family, talent, and altruism. Whether linked by a contestant, a region, or specifically an American identity, the collective identities shared by *American Idol* fan groups reflected the *American Idol* ideals of family, congeniality, and philanthropy.

Chapter 8

Collective Identities On the *American Idol* Message Boards

Reality TV was first among other genres of television programming in one important way: interactivity. Several reality-TV programs embrace convergence-enabled interactivity, encouraging audience members to vote through text messages and participate in online forums (Andrejevic, 2004). While many TV shows have their own dedicated online forums, reality TV also includes interactive voting. In the summer of 2000, *Big Brother (UK)* was the first reality-TV program to encourage audience participation. During that season, the *Big Brother* website received over three million hits daily (Tincknell and Raghuram, 2004). Audiences were able to watch live feeds of the cast online, twenty-four hours a day. This exemplified an older medium (television) supporting and supported by new media (the Internet), and also illustrated how the convergence, or blending, of new and old media encourages interactivity between the audience and the show. In their discussion of *Big Brother*, Tincknell and Raghuram (2004, p. 254) attributed its success to the "range of sites of access" that audiences had to the program. Web cams and chat rooms facilitated audience participation. Moreover, the "webcam images offered a convincing sense of immediacy and 'liveness'" (p. 255). (Although audiences may never know if the scenes really were live.) Audiences also crucially participated by voting contestants off the program and choosing the tasks house members had to perform (Tincknell and Raghuram, 2004).

American Idol is similar to *Big Brother* in that both ask the audience to vote and both accumulated high ratings. Audience members vote for their favorite contestants through their cell phones, interactively linking the audience to the show (Andrejevic, 2004). Aside from voting, audiences may participate in

This chapter appears with courtesy of Cambridge Scholars Publishing.

online discussions. As is true with other television genres, audiences are also able to watch interviews and ancillary show material online. This interactivity connects the audiences and the program, fostering an emotional link.

In Jenkins' (2001, p. 93) discussion of cultural convergence, he noted, "media convergence fosters a new participatory folk culture by giving average people the tools to archive, annotate, appropriate and recirculate content." As people use many media simultaneously, culture becomes more homogenous across channels. Participation is a key element in the convergence of computers and television—interactivity encourages a form of audience agency. Audiences can participate through online feedback on discussion boards, voting for participants of reality shows, entering online contests, and deciding when to watch TV through digital video recording. This audience participation is a "way of getting *American Idol* viewers more deeply invested, shoring up their loyalty to the franchise and its sponsors" (Jenkins, 2006, p. 70). Indeed, the process of voting weekly increases the audience members' engagement (Jenkins, 2006). The online communities enhance show loyalty, as well as act as a focus group for the show producers (Andrejevic, 2008).

Kavka and West (2004, p. 138) also asserted that reality TV is a social technology: "People watch it all the time, are affected by it, and incorporate it into the structure and reference of their daily lives" The global amount and variety of reality-TV programming suggests, "that reality-TV audiences seek, through television consumption, a greater sense of community and even agency" (p. 138). Thus, reality-TV consumption is often not passive. Some audience members choosing to watch reality TV also may choose to exert agency through interaction such as voting or online discussion boards. The immediacy of interaction produces social intimacy, further linking audiences to the program (Kavka and West, 2004). Thus, the cultural studies notion of the active audience becomes the interactive audience, literally negotiating meaning through choice (Tincknell and Raghuram, 2004). The hegemonic ideological meaning may be undermined through the audience authorship of the program (Tincknell and Raghuram, 2004) or made stronger through the illusion of audience authorship. Nonetheless, the dominant meaning may also be reinforced through audience acceptance. After all, producers still do have control over editing and production of the show itself. While participation begets interactivity, altering the relationship between TV and viewers, audiences still are required to negotiate meanings constructed by the producers. Analogous to *Big Brother*, *American Idol* is another version of interactive convergence and audience.

AMERICAN IDOL: INTERACTIVITY

American Idol embodies the new convergence of technologies into one cohesive entity. At first glance, it is a product of traditional media, a television program. Upon further examination, the participation facilitated by the convergence of technology, notably the ability to vote via text messaging, the online contests, and the online feedback forums, is representative of a new genre of media products. *American Idol* fans, by watching and participating, support the show itself and the individual contestants competing. The show fosters online fan groups, continually exhorting the audience to visit the official message boards as well as asking the audience members to participate through voting.

Audience voting participation is a means of exerting some agency and creating community. During the 2004 season of *American Idol*, 13.5 million text-message votes were received. Forty percent of the people sending in votes said they had never text messaged before voting (Stakutis and Webster, 2005). The 2006 season generated over 65 million text messages, according to FremantleMedia Vice President of Integrated Marketing and Interactive-Americas Keith Hindle (Atkinson, 2007). The April 25, 2007, special episode, *American Idol Gives Back*, spawned over 70 million text-message votes. For each vote, an undisclosed amount of money was donated to charity (American Idol, 2007). In 2008, a Nielson report found that the average *American Idol* audience participant voted via text message thirty-eight times just in the month of April (The Nielsen Company). For television audiences, participation may instill a feeling of belonging in a show's audience, fostering a community and enhancing the relationship between a viewer and an entertainment production.

American Idol took *Big Brother's* concept of interactivity farther. *Big Brother*, the initial modern television production to encourage viewer voting, asked viewers to vote for their least favorite contestant, resulting in his or her leaving the program. Similarly, a sizeable number of people use the *American Idol* discussion boards to air thoughts and debate contestants' qualities, as well as deliberate a variety of issues (FOX Interactive Media, April 18, 2007). Additionally, through the *American Idol* website, viewers can start their own *American Idol* blog, "Get the Gear!" (purchase ancillary products), or "Play the Game!" (FOX Interactive Media, April 18, 2007). Other venues of interaction are contests, such as the *American Idol* songwriter competition. At the start of the 2007 season, the *American Idol* website solicited audience members to submit original songs. The 2007 *American Idol* champion, Jordin Sparks, performed the winning song, chosen by online voters. The profits of the publishing contract for the song are potentially great (FOX Interactive Media, April 18, 2007). This contest occurred in the following season as well. In another example of audience interaction, show host Ryan Seacrest urged viewers to add *American Idol Gives Back* as a MySpace friend. MySpace is an online social networking website owned by NewsCorp., which also owns FOX, the network on which *American Idol* is broadcast.

The interactivity of *American Idol* has proliferated, converging the functions of television, telephony, and computers through audience participation. Considering the previous observations of *American Idol's* popularity and interactivity, it can safely be assumed that *American Idol* offers audience members high-level engagement through various media. Often, as audience members engage with shows, they become a community with a particular identity.

MESSAGE BOARD DISCOURSE ANALYSIS

American Idol collective identities formed through comments on the *American Idol* message boards. This chapter uses Tajfel's (1984) social identity theory to investigate how group collective identities formed within the 2008 *American Idol* message boards. On the *American Idol* message boards, two types of collective identities were primarily visible and are explored here: the first is associated with a particular contestant, while the second is associated with being a fan of the show. The collective identities allow group members to identify not only with the group, but with the show as well, encouraging commitment to watching the show and upping potential consumption of the show and affiliated products. Jenkins (2006) asserted that fan engagement links audience members to the show and contestants as brands. Accordingly, collective identities engendered by community membership create a stronger link for participants to the show itself. Not only are participants community members, possibly gaining pleasure from participating, but pieces of their identities are also bound to the show. Additionally, communities based around television shows help to construct the "material and social" conventions of American society (Andrejevic, 2008, p. 43).

Several studies have examined the online fan activities of various TV programs, including Clerc's (1996) study of *X-Files* message boards, Baym's (1998) study of soap-opera fans, Pullen's (2000) study of *Xena: Warrior Princess* fan sites, Foster's (2004) study of *Survivor* fan boards, and Andrejevic's (2008) study of the website *Television Without Pity*'s message boards. Additionally, Jenkins (2006) found markers of consumer loyalty and causes of consumer distaste on the seasons two and three *American Idol* message boards. This study also found that gossip, or social currency, was a reason to watch. It follows, then, that fans continue to make meaning through participation online, constructing and circulating a fan culture based around a program (Foster, 2004). Jenkins (1988), writing before the popularization of the Internet, asserted that fans not only watch shows, but also are emotionally invested in a program, as evidenced in discussions about the text and by partaking in communities. The invested collective identities found here correspond with brand communities. Brand communities, a marketing term, are groups of people who connect to and through a product, and are highly loyal consumers (Jenkins, 2006). The *American Idol* collective identities somewhat match this definition, although typically past brand communities were centered not upon media items, but rather products. However, as Foster (2004) noted, discussion boards are often created and maintained by a program's production team, indicating the hope for increased consumer interactivity and consequently loyalty. This "structured interactivity" is created by the program's team and designed to develop fan culture (Brooker, 2002, p. 325). While the design by the program's producers cannot be denied, the discussion and debates on the official message boards often provide important insights into the thriving fan culture (Foster, 2004).

During the 2008 season of *American Idol*, I lurked (a common online term that denotes observing without participating) at least four hours per week in the discussion areas of the official *American Idol* message boards, collecting transcripts of online discussions. Each season of the show has a dedicated message board on americanidol.com, with topic folders dedicated to individual contestants, as well as general *American Idol* discussion, rules and FAQ (frequently asked questions), and other categories that change over time. Within each online folder there may be thousands of discussion threads. However, after a contestant was eliminated, the activity in his or her dedicated folder slowed. It was impossible to gauge how many people posted to these forums—there were most likely several thousand, with about two hundred posting on an almost daily basis. This analysis focused primarily on the most active threads and noted any indicators of community formation, maintenance, or termination. In order to discern patterns and themes, the investigation searched for discourse cues including repetition, exaggeration, and positioning of fan in the community through word selection. Additionally, the investigation included notice of like or dislike for particular contestants. Aversion to certain contestants or collective identities exhibited the us-versus-them social identity theory (Tajfel, 1984) phenomenon of reinforcing community identity through common distaste. In keeping with the key components of social identity theory, the discourse analysis particularly noted manifestations of categorization, identification, and comparison of groups.

COLLECTIVE IDENTITIES

There were two primary types of collective identities present within the 2008 *American Idol* message boards: a large overarching identity as fans of *American Idol* and smaller collective identities based around each contestant competing on the show. First, this chapter explores distinctive collective identities constructed around individual contestants, visible through discourse about the groups, including self-identification, inside jokes, relationships, attribution of qualities to the community, and favorable comparison with other communities. Next, the overall *American Idol* fan collective identity is considered. This identity was demonstrated through discourse emphasizing fan unity and respect, relationships built online that transcend seasons, and shared empathy between contestant groups.

Collective Identities Around Individual Contestants

As explicated earlier, social identity theory is the process of categorizing, identifying, and comparing social groups (Tajfel and Turner, 1986). This process was apparent in the smaller social groups that relate to individual contestants. Social groups formed around shared interest in a contestant, leading to a collective identity based on group membership. The group members favorably compared their group with other groups.

Fans categorized themselves according to contestant favorites. At the beginning of the season, fan group members communally created one or two names associated with their chosen contestant and referred to themselves as that name throughout the season. The names were chosen through a group process of suggestion, discussion, and agreement. Every group associated with a top-twelve contestant underwent this process. For example, members of David Archuleta's group were the Archies, Carly Smithson's group called themselves the Fighting Irish, and members of Michael Johns's group, the Aussie Posse. David Cook's fans anointed themselves the Wordnerds, referring to David's passion for crossword puzzles, and also, familiarly, Cookies, for his last name. Once the groups were established, aided by the folders dedicated to each contestant, it was very common to see a member self-identify, saying, "I'm a cookie, too" as Jpink did on April 11, 2008, in the David Cook folder. They also greeted each other as fellow group members, such as when Digileach wrote, "Good morning Word Nerds!" (April 17, 2008). Each group had a discrete identity (which usually continues to exist even after the season ends), no doubt facilitated by the website's individual folders dedicated to each contestant. The program provided an official space for the communities to thrive, prompting their creation and perpetuation.

During the subsequent 2009 season, the group names decided by the fans appeared officially sanctioned, appearing on a section of the website dedicated to official lists of group members. These separate collective identities were within the overall collective identity as *American Idol* fans, yet nevertheless distinct from each other.

Discourse cultivated and encouraged the formation of the contestant-centered communities. Inside jokes often began on one thread and then migrated to another. If a poster did not read the original joke, he or she had no idea what the discussion was about, and members were not apt to tell newcomers. Those who recognized the joke may have felt essential to the community. Games were another community-building exercise that occurred at least once a week in every single contestant's folder. For instance, in one thread or conversation, the participants played a game where each contributed three words, so that everyone jointly created a story. This connected people through a communal activity and shared humor. In another example of an activity, members created an imaginary hotel, with each poster describing how he or she decorated his or her room. Similarly, they threw imaginary parties and each person informed the others what he or she was bringing to the party. These games built relationships and helped develop the social identity of the community.

Thus, within each community, relationships were built over time through discussion. Posters became familiar to each other and friendships formed through shared interests. In an interesting example of friendship and group identity, during a segment that aired live on the program, board user and Wordnerd Mallory09 was selected to ask contestant David Cook a question via telephone. She asked if he was single. That night, the boards were abuzz with people congratulating Mallory09 for getting on the air, asking a question everyone seemed to want to know the answer to, and for ably representing them. Kscook detailed her experience watching the show, saying, "I was screaming and I grabbed my husband and said, 'I KNOW HER!!!'" (April 16, 2008). The Wordnerds were excited because they identified Mallory09 as an integral part of their group and as a friend. DoctorBice congratulated her, saying,

> I knew it had to be you! When I heard Mallory, I said YES!!!!!!!!!!!!! Only thing—I wish you should have given a shout out from the Word Nerds, but I know your time was limited. I guess you covered us when you said 'all us ladies.' Great job!!!!!!!!!! (April 16, 2008)

All the Wordnerds rallied around Mallory09, reveling in her success and applauding her composure on the air as she actually spoke to David Cook. Crzy4DavidC said, "I am so happy that you did this for us Mallory!"

(April 16, 2008). Mallory09 was a member of the group, and her successful involvement in the show represented an achievement for everyone within it.

Mallory09's experience was an exceptional example of a member representing the group and the relationships within it. However, once relationships existed, the members of the communities frequently rallied around each other, offering condolences, advice, and emotional support when necessary. When Wordnerd gigibeans reported to the board that she had lost her job, she immediately received support. Lovincook wrote about getting a new job, "You should be able to take your pick . . . find the best job available . . . don't settle. Anyone would be PROUD and LUCKY to have you a part of their team" (April 17, 2008). The members offered friendship and encouragement to each other, just the same as off-line friends. The relationships among the members built bonds between them. In turn, those bonds established a social group and subsequently a collective identity that the members shared.

Once groups categorized and self-identified, they compared themselves to other groups. Other groups also strengthened the identity of each group through recognition. On performance nights, a noticeable pattern emerged of group members visiting the folders of the other groups and wishing each other luck, establishing a basis for comparison. For example, on April 29, 2008, in the David Archuleta folder, a David Cook fan started a thread entitled "Good Luck from the Word Nerds!!" Within it, Wordnerds wished David Archuleta and the Archies good luck for the evening's performance. Fan Rockview posted: "Good luck to David A. and the Archies tonight. Hoping to see a David A. and David C. finale!" (April 29, 2008). This occurred repeatedly in each of the remaining five contestants' folders on that day, acknowledging the other groups' existence and indicating membership for the poster.

Members in one group often referred to other groups by name, such as when brittneybabyy wrote in the David Archuleta folder: "I bet the Word Nerds are having a fiesta. Ahah. But I don't mind—since I do love David Cook! But our little David was still good! Let's just keep those votes coming. The full 2 hours people!" (March 25, 2008). Each community noted the existence of the other and attempted to favorably compare themselves. This established existence and presumably boosted members' self-esteem.

Each group attributed qualities to themselves and other groups as a measure of evaluation. For example, Lavalamp77, a member of contestant Syesha Mercado's fan group named Syesha's Syclones wrote, "We may not be as rabid as other contestant fan bases on the web, but at least we know talent" (April 3, 2008). According to Lavalamp77, Syclones differentiated themselves through their higher regard for aptitude, a positive distinction from the other groups. In another example from the Syesha group, Nostradamus2k described the Syclones as a "fan base that doesn't go around belittling/

degrading other contestants, but instead directs their energy into constructive support of her performance efforts" (April 3, 2008). This poster identified the collective identity of the fan base and favorably positioned the Syclones against other groups.

Since an integral part of favorable comparisons between groups was how fans understood their own group, the assumed reputation of each group was important. Members cared about the image of their group. Gigibeans posted:

> The Word Nerds are starting to get a bad reputation on these boards. I've noticed (lately) that a few Word Nerds are bashing other contestants. I've also noticed that quite a few Word Nerds get a little overly defensive at times (including myself) and this has got to STOP!!! We need to regroup and once again become the positive-friendly fan base that I know we are. (March 21, 2008)

This statement once more asserted that the Wordnerds were a distinct group through indication of a collective identity. It also demonstrated concern for how other groups viewed the Wordnerds, showing a search for positive comparison. If it was known that the Wordnerds were a negative group, positive comparisons may have been difficult. Members wanted to be able to advantageously compare their group.

Through application of social identity theory to the 2008 message boards, it is evident that social groups and consequently collective identities formed around the individual contestants. These were continued after the season ended, during the contestants' summer concert tour. Each fan group decided a representative color and agreed to wear it to the shows. No two fan groups chose the same color. For example, contestant Brooke White's fans agreed to wear yellow, contestant Jason Castro's agreed to wear baby blue, and contestant Chikezie's fans agreed to wear gray. Even after the show is over, the contestant-based collective identities remain.

OVERALL *AMERICAN IDOL* FAN COLLECTIVE IDENTITY

In addition, an overall fan identity was apparent. While the distinct fan communities doubtlessly existed, an overarching community composed of the varied contestant communities worked as an umbrella, encompassing all message board participants. The inclusive community functioned to provide a home base for participants, furnishing friendships and a place to comfortably share interest in the show and opinions about it. *American Idol* community membership also supplied a portion of self-identity for members. General fan identity characteristics included discourse stressing fan harmony, connections beyond one season, and compassion for varied contestant groups.

A key indicator of the general fan collective identity was the frequent discussion comparing contestants from various years. For example, fans compared 2008 contestant Kristy Lee Cook to 2006 contestant Kellie Pickler and 2005 contestant Carrie Underwood. David Archuleta, a 2008 contestant, was compared with 2003 contestant Clay Aiken. This indicated fanship of the show in general, as well as of contestants individually. Message board debates about contestants' attributes from various seasons illustrated that a larger community had existed for years. Similarly, demonstrations of knowledge about *American Idol* pointed to a general fandom of the show and the existence of the overall collective identity. Through the message boards, fans formed an imagined community, sharing a communal identity (Anderson, 1983/2003). The imagined community was apparent through familiar greetings by name, references to previous conversations or events, and enjoyment of common interests, although most members most likely never actually met (or will meet) each other in person. This collective identity continued across space through the Internet and across time through the various seasons of the show.

On the program, the show frequently used a family metaphor to describe the relationships of the contestants and people working on the show. The family metaphor of the contestants carried over to the fan boards, where a communal feeling was prevalent. According to Foster (2004), the *Survivor* online activity imitated the nature of the game, with online participants one-upping each other and vying for status. In a parallel, the *American Idol* message boards reflected the familial and congenial nature of the majority of the show.

McFireMedic, a member of contestant Kristy Lee Cook's fan group Kristy's Krushers, posted on a thread, titled "Welcome new members and fans," that

> You will find that we treat all of you, and each other as if Kristy was reading this page. We conduct ourselves that would make her proud of her fans (sic). We ignore bashers. And we do not go to other fan bases and start trouble, we are 99 percent bash free. Nor do we post negatives about the other Idols. (April 10, 2008)

Like the other fan groups, the Krushers prided themselves on the civility and respect of their group. The online fan groups' behaviors mirrored the overall tone of the show, amiable and good-natured. When a competitor was eliminated from the show, the show tone was friendly, with the contestants professing their friendship and fondness for each other, proclaiming how much they would miss everyone and thanking fans. This feeling was replicated on the message boards. Flaming (starting fights) was looked down upon. Instead, commenters worked to facilitate a nice place, welcoming newcomers and advocating for cordial relations between groups. As the end of the 2008 competition drew near, bambideerjean posted in the David Archuleta folder about the two finalists, David Archuleta and David Cook, "I think they are both rooting for each other, you can see the closeness with these two and I totally agree why can't these two fan bases pull together for our Davids? It's a shame that there can't be 2 top spots. IMO they are both #1" (April 27, 2008). The good relationship of the two Davids was mirrored on the message boards, highlighting the overall communal identity despite the existence of two factions.

While smaller social groups forming collective identities did exist, relationships that lasted throughout several seasons transcended these smaller groups. Relationships built online over shared interests, such as general interest in the show and individual contestants on the show, were evidenced through greeting members by name, referencing conversations from other years, and reminiscing. For example, Divinerev1 wrote on contestant Syesha Mercado's message board:

> Hello everyone, I'm not new to AI boards, but I have not been on since last season. To make a long story short, I was flipping through with the remote one night and saw Syesha singing and I was very impressed by her. I'd like to throw in my support for her now if it's not too late. Oh yeah, big shoutout to my girly girls, Sis and Idolchicklet!! MUAH! I'm back, ya'll! (April 1, 2008)

Poster sistersledge warmly greeted Divinerev1, saying "And back attcha! Rev!! I felt like break-dancing when I saw your name." Cats chimed in, "I REMEMBER YOU! Lakisha Divas!" (April 1, 2008). They were long lost friends, reuniting on the *American Idol* message boards. "Lakisha divas" referenced a contestant from the year before, suggesting a previous relationship. Communities were built upon these relationships, forming the *American Idol* fan collective identity. The members may have originally met and forged a friendship in the season-six message boards, but continued online friendships during season seven. They belonged to an individual contestant's fan group, but in general belonged to the overall community as well, as indicated by the relationship that transcended contestants.

Additionally, when a contestant was eliminated from the program, the various communities commiserated with each other. This occurred every time a top-twelve contestant was eliminated. Posters empathized with other members, providing an overall feeling of community, linking the various groups together. For example, immediately after contestant Michael Johns was surprisingly eliminated, a thread titled "This Word Nerd Is Heartbroken" was posted in the Michael Johns folder. AmariesIdol posted on Aril 11, 2008:

> After the show last night, I finished crying, took a deep breath, and told myself not to worry because we haven't seen the last of Michael. . . . He may have left the show, but he hasn't left our hearts. He's going to be just fine, and he has all of us here to support him through what I am sure will be a shining career!

Everyone banded together to support the eliminated contestant, Wordnerds and Aussie Posse (Johns's fan group) members alike. Similarly, a fan with the handle davidcookislove wrote: "Again, Aussie's (sic) I am SO sorry and we word nerds are here whenever you need us =)" (April 11, 2008). Davidcookislove identified as a Wordnerd, yet shared grief with the Michael Johns fans. This indicated that individual fan groups based around contestants clearly separately existed. Simultaneously, a larger *American Idol* fan collective identity existed as well. Many fans, no matter which group they identified with, commiserated with the other fan groups. While social identity theory indicates that Wordnerds may have favorably compared themselves to the Aussie Posse, as David Cook was not eliminated (and indeed some fans may have been rejoicing as the competition dwindled), instead, they publically sympathized. This indicated a membership in the larger community that at times may have superseded contestant group membership. On the *American Idol* fan boards, when a contestant was voted off, the dominant emotion was sadness or, when deserved, ambivalence. It was not triumph or gloating over the competition's failure. Success was earned through making it past each round, and not by denigrating competitors, who were supposedly more like family than foes. This sympathy reflected the existence of an overall collective identity as general *American Idol* fans.

Other fan groups expressed sympathy for the Aussie Posse too. Fanofidol99 wrote, "We Archies are really upset. He was robbed and it sucks big time" about Michael Johns being voted off (April 10, 2008). The Archies were noted as their own group by the poster, and empathy for the Aussie Posse was present as well. Both fan groups undeniably existed, but the sympathy indicated a bond between the two groups and the communal nature of the overall fan group. Still another group wanted to console the Aussie Posse—Jason Castro's fans, the Dreadheads. CastroWordNerd started a thread titled "Hugs from a Dreadhead," explained her sympathy for the Aussies, and ended her post with "CHEERS TO THE AUSSIE POSSE!!!!!" (April 11, 2008). On the same thread, jennb78 added, "The dreadheads love you guys!" (April 11, 2008) All of the remaining seven contestant groups rallied around the Aussie

Posse, expressing sympathy, pity, and compassion, reinforcing the overall group mentality of the collective fan social group.

Social identity theory posits that groups compare themselves to other groups. And while they clearly did compare themselves, an overall camaraderie was plainly present and desired. Msdog, in the David Cook folder, noted her individual community as well as the importance of the overall community:

> I was really sad to read on the general discussion board that David Cook fans were some of the meanest. And one person stated that one particular David Cook person was the meanest fan on the entire AI board! Eek! Remember we are *American Idol* fans not *American Idol* gladiators or alligators. (March 24, 2008)

Overall, all fans belonged to the same all-encompassing community, as well as their smaller contestant-based groups.

CONCLUSION

Through the lens of social identity theory, two collective identities were observable in the 2008 *American Idol* message boards: a comprehensively inclusive fan collective identity and smaller contestant-based collective identities. The interactivity of *American Idol*, particularly voting and participating on the message boards, created a new type of audience member who was often actively involved with the show and emotionally connected to the show community.

This connection with *American Idol* social groups led some audience members to self-identify in part through the program and social group. Social groups initially formed through categorization, identification, and comparison between groups. These social groups advanced into collective identities, with participants actively identifying as Archies or Wordnerds or Dreadheads, for example. Once the contestant a group was based around was voted off of the show, the individual group members would often declare their allegiance to another contestant, indicating their membership in the overall fan group as well.

American Idol facilitated the formation and persistence of these groups by the creation of the forum; their existence is beneficial for the show. Within these separate fan collective identities, relationships constructed the bonds supporting the communities. Emotional connections may have existed between the fans and their chosen favorite contestant, between the fans and the show, or among the fans themselves. Regardless, the fan communities functioned to support the contestant, to support the community itself, and to support the popularity of the show. Participation and identification with the fan communities resulted in self-identification with collective identities associated with the program. Identification with a community associated with *American Idol* ensured audience loyalty. These communities intensely connected individuals to the program and also increased brand loyalty (Jenkins, 2006). In addition, the collective identities cultivated by the fan message boards helped to assure the continued popularity of *American Idol*, as well as the social and material status quo it supported.

American Idol engenders community both through the show discourse and online, which contrasts with the show's discursive emphasis of individual achievement. Both community and individualism function to support the economic and cultural status quo. Individualism endorses the ideal of equality, that everyone can achieve the American dream of financial success. The *American Idol* communities function to support the show's popularity and revenue streams. Thus, while seemingly oppositional, both components of the individualism/community duality serve the same function; to preserve the extant economic and cultural systems that facilitate *American Idol's* success.

Chapter 9

American Idol Press Coverage: 2002 and 2008

This chapter examines some of the themes, patterns, and general overtones in the press coverage of *American Idol*. News contains narratives, characterizations, and mythology. In particular, the role of news in producing and expressing an American mythology is pertinent to this book. This chapter contains two discourse analyses of *American Idol* news coverage: one of the 2002 first season and a second of the 2008 seventh season. I chose these two particular years as they bookend the larger discourse analysis described earlier. Finally, this chapter compares the findings of both analyses.

Analogous to entertainment television, news helps form and maintain culture. Within news, narratives help audiences construct a believable reality, through repetition and familiarity (Barthes, 1977; Zelizer, 1997). Within these narratives, recognizable cultural codes impart meaning about daily life as well as societal ideals. Therefore, in addition to relating facts, news stories convey how to understand stories (Bird and Dardenne, 1997).

Tuchman (1978) noted that news is inherently ideological. As the intrinsic narratives repeat, they combine to form an ideology. In news and in daily life, meanings are understood through the lens of cultural dominant ideology (Williams, 1981). Correspondingly, news promotes an understanding of one version of a story, linked to the dominant paradigm of culture. Like television, news stories function to transmit, support, and maintain dominant ideologies. Studying news associated with *American Idol* permits further insight into ideological codes associated with it.

PRESS COVERAGE STUDIES

The studies included articles from newspapers such as *The New York Times*, *The Washington Post*, *The St. Petersburg Times,* and *The Philadelphia Inquirer*. Magazines in this study included *Variety*, *People*, *Billboard*, *Advertising Age*, and *Newsweek*, as well as others, and television coverage included programs such as *The Today Show* and *Larry King Live*. Radio coverage included shows broadcast on National Public Radio. This press coverage included trend pieces, reviews, programming announcements, and commentary. The functions of the pieces varied. For example, morning talk-show interviews provided publicity, while commentary and reviews published criticism. These different types of journalism combine to form an overall news discourse.

In the discourse analysis of the 2002 press coverage, the themes concerned judge Simon Cowell, the economics of the show, the contestants' characters, and a rags-to-riches narrative. In addition, this study noted two overarching attitudes toward the show: surprise and general derision. The 2008 seventh season discourse analysis revealed themes of economics, past contestants, current contestants, specific contestant Carly Smithson, and authenticity versus artificiality. The 2008 press coverage lacked the awe and disdain overtones found throughout the 2002 press coverage. By 2008, *American Idol* and its immense success were no longer cause for surprise or disparagement.

2002 PRESS COVERAGE ANALYSIS

For the first discourse analysis, Lexis Nexis returned 1,136 articles containing the words *American Idol* appearing between June 1, 2002, and September 30, 2002. Articles or transcripts concerned primarily with ratings or recaps, or those that only mentioned *American Idol* and were not about the show were discarded. Approximately 177 media pieces were discarded, equaling 15 percent of the total number. Thus, 959 articles and transcripts are included in this discourse analysis.

The discourse analysis consisted first of reading all the articles, including radio and television transcripts. Next, I sorted and identified patterns within the articles and transcripts. This process discovered four themes: Simon Cowell and his caustic opinions, economics including a general discussion of reality television and *American Idol's* place in the genre, contestants' characters, and an American dream/rags-to-riches theme. Moreover, two overtones were present throughout the press coverage: derision and awe. Reporters mocked the show, particularly its seemingly lowbrow appeal. Despite this contemptuous insinuation, the other general connotation was astonishment and wonder at the popularity the show achieved, as well as the convention-defying nature of various show elements, such as Simon Cowell's biting criticism or the program's blatant product placement.

Simon "Mr. Mean" Cowell

A prominent theme was marvel and shock at judge Simon Cowell's patently rude criticism. Someone like him had truly never been seen on American television (except perhaps for Anne Robinson, host of the short-lived game show *The Weakest Link*, whom Simon protested he was nothing like). That first season Simon criticized contestants' singing ability, appearances, and personalities, often in a cruel manner. The novelty of this, along with his physical attractiveness, seems to have enthralled media and audiences alike. While certainly Simon's rude comments were a ploy to gain attention and viewers, this news discourse analysis reveals his comments and presence helped support the American ideals present within the show.

One American ideal is that anyone can succeed through talent and hard work. The news coverage insinuated that Simon's presence on the show supports the idea of talent meriting fame. According to Simon, "It's sometimes fairer to tell someone the truth" than to prolong his or her attempt at a music career through false praise. He claimed this was his reasoning behind comments said to auditioners, such as: "That was dreadful. And I'm saying that to be kind, because you will never, ever have a career in singing"; "There are only so many words I can drag out of my vocabulary to say how awful that was"; and "You actually sing like a train going off the rails" (Battaglio, 2002a).

Simon's endorsement was made more valuable by his quickness to reject people. Thus, this connotes the idea that true talent is rewarded—the notion that being in the right place at the right time with the right skills can create stardom. If an auditioner was actually talented, Simon's praise could elevate a contestant into a front-runner for fame. As for poor singers, as Simon said to Diane Sawyer, "I actually think I'm doing most of these people a favor by telling them, which they've obviously never been told in their lives, 'you can't sing'" (Brune and Ross, 2002a).

Second, Simon's British accent, vocabulary, and brisk mannerisms reiterated in the press were a clear counterpoint to his American co-judges. His un-American nationality may have added to his exotic attractiveness. Simon is unambiguously not an American. Therefore, his rudeness was foreign and could not be attributed to American culture. His British nationality and bad attitude pointed out in the press may have further solidified oppositional collective identities, such as an American national identity. Americans like Paula Abdul and Randy Jackson were seemingly polite, often speaking banal niceties. Simon, a caustic Brit, was a foreigner, pointing out the kindness of Americans through comparison.

Simon's scathing comments, often discussed in media, displayed another cultural component. Simon flouted norms, saying exactly what he thought, disregarding civility standards. Randy Jackson pointed out American cultural practices on the *Today* show, disagreeing with Simon's method of rejection:

> It's just that, you know, you can't call people losers. I think this is America and I think there's a right way and a wrong way to do things. We're not here to bust dreams, we're not here to destroy dreams. I agree with him, if somebody's really bad, they should know that because the entertainment business is very brutal and very honest, and that's what's great about this show. But it just, you know, it just got to me at that point, I was just like, 'come on, you can't call people losers.' (Wald, 2002b)

American conventions dictate that being overtly rude is unacceptable, a convention that Simon did not follow. As Simon pointed out about Americans, "I think you're more polite than the English" (Keveney, 2002a).

News coverage often cited a related aspect, *schadenfreude*, or delight in another person's suffering. Simon's criticism encouraged this emotion. *American Idol* has only one winner and a great number of losers. The popularity of *American Idol* partially attests that audiences seem to enjoy watching other people fail at least as much as they enjoy watching someone win. Similarly, programs like *The Gong Show* and *Showtime at the Apollo* echo a similar sentiment, wherein people not good enough are humiliated, emotionally hurt, and thrown off the show. As one writer noted about *American Idol*, "It's great fun, if you enjoy other people's misery" (DeLuca, 2002). Journalists' shocked reactions to Simon's behavior implicitly revealed an American cultural norm of civility (at least as demonstrated on television).

Character of Contestants

A more amiable human-interest theme concerning contestants' characters provided a counterpoint to the Simon-centered discourse. This coverage introduced the contestants and provided background information to prospective and current viewers. Knowledge of the contestants may have enticed watchers to continue viewing as well as attracted new audience members. Whereas media reinforced Simon's depiction as the cruel judge, the top-ten contestants were portrayed as overly nice, moral people deserving of the recording contract prize. The top-ten contestants were all described in terms of their families, jobs, and physical appearance. This made the contestants seem average.

Most noticeably, each of the top ten was humanized and given a single distinguishing feature. These features were mostly average attributes, normalizing the televised contestants. For example, Nikki McKibbin's single-mother status was frequently mentioned. Similarly, Christina Christian's life as a college student and Kelly Clarkson's former job as a cocktail waitress made each seem normal and down to earth.

The press coverage portrayed all of the contestants as fundamentally "nice." Contestants' parents also appeared in the press, contributing to the establishment of each competitor's character. Glen Helton, contestant R.J. Helton's father, appeared on *CNN Live.* He called his son a "special kid" and "humble" while also claiming that R.J. had "international appeal" (Grady, 2002). Justin Guarini's mother also thought her son was "a special kid . . . with a very even keel" (Waxman, 2002). Front-runner Justin Guarini had several pieces written about him, detailing facts about his background growing up, experiences with music, and kind nature. News coverage noted that Justin's popularity was propagated by his fans and his appearance: "The show has been drawing a disproportionate number of preteen and teenage girl viewers—making 23-year-old finalist Justin Guarini, he of the drowsy brown eyes and flop-mop hair, the undisputed favorite" (Waxman, 2002). In the ultimate nice-guy move, after Kelly was announced the winner, Justin confessed he had been wishing for Kelly to win because "it just felt so right" for her to win and "she did do better" (Friedman, 2002b).

This theme established that each contestant was a viable candidate based on his or her integrity or moral fiber. Each was praised for his or her human qualities and personality, and not primarily for singing ability. This helped add to the conception that the contestants were all average people. Additionally, each was also stressed as hardworking, wholesome, and deserving of winning.

Economics

A third theme throughout the 2002 *American Idol* press coverage was economics. At the time of its debut, it was one of the first shows to blatantly utilize product placement; Coca-Cola and Ford were prominent throughout every episode. This was reported with a mixture of disgust and awe. In addition, various other ways *American Idol* earned income were explained, such as contestant record deals. A further component of this theme was the rising visibility of reality television and *American Idol's* place within the genre.

Prior to the airing of *American Idol*, most product placement was intended to be surreptitious; it occurred, but audiences were not supposed to consciously notice the products. Products were mentioned or used diegetically within programs. Like other reality shows, *American Idol* disregarded this standard. *Variety* reported that,

> Fox has also sealed a multiplatform deal with Coca-Cola that includes media time and branded elements within *American Idol*. Show includes elements such as 'Coca-Cola Moments' and a 'Coca-Cola VIP Red Room.' As part of the deal, Coke will produce two national radio promotions for *American Idol*, sponsors the show's Web site and has an option to sponsor an eventual CD and concert tour. (Schneider, 2002)

Product placement did not occur as frequently in scripted programs as in reality television for an assortment of reasons. Initially, executives assumed that product placement would discourage companies from buying commercial time. Additionally, working product names into scripted sitcoms and dramas is difficult. Reality television and product placement provide a simple way to avoid the clutter of a crowded media space and negates the fast-forwarding of commercials through the use of digital video recorders (Battaglio, 2002b).

By the end of the 2002 *American Idol* season, the ratings were phenomenally high. Those who had originally invested with the program were surely pleased. As *USA Today* reported, "Coke got in on the ground floor in March with *American Idol* for less than $10 million. That was a bargain based on ratings that were off the charts for the 12-week program, beating network promises by about 10 percent and capturing 23 million viewers for the finale" (Howard, 2002). Coca-Cola was prominently featured throughout the set of *American Idol*. There was the Coca-Cola-label-saturated red room (instead of the traditional green room) where contestants were interviewed on camera and waited for their turn onstage. Additionally, several Coca-Cola insignias were built into the decor, including a Coca-Cola branded couch, photo booth, and pinball machine. Contestants and judges drank from conspicuously labeled Coca-Cola cups. This overt sponsorship was an example of "how products are being integrated into reality programs so viewers can't avoid them" (Battaglio, 2002b).

The contestants' recording deals were another fresh innovation that positively affected the program's economics. One article noted that "industry insiders say the show exemplifies a winning marketing strategy: the same people who vote for the winner are also likely to buy the singer's records" (Hay, 2002a). Not only was the audience doing double duty as fans both of the show and the winner, audience members were considered "vote-casting viewers acting as talent scouts" since they chose the winner (Gundersen,

2002). Many articles pointed out Simon Cowell's vested interest in the winner as well, reporting that the winner would be signed to an RCA record label with which he was affiliated.

Reporters enumerated and admired the many platforms designed to appeal to audiences, such as television, albums, and voting mechanisms. As one article noted, "'I envisioned all along that this show would be a multimedia phenomenon,' says Simon Fuller, creator/executive producer of *American Idol* and the founder of the London-based 19 Entertainment Group, which includes a management company, production company, and record label" (Hay, 2002b). *American Idol's* many media facets were unusual and inventive. Even before the season ended, plans included the winner's album, a compilation album, a nationwide tour, a DVD of the series, and a sequel to the series. For a fan of the show, there were myriad opportunities to see and hear the contestants. These activities also kept the show in the public eye until the next series premièred in February 2003 (Keveney, 2002b). The overall tone of the economic press coverage simultaneously contained wonder and dread at the program's impressive and innovative financial components.

American Idol was also discussed in terms of the reality television genre. This discussion included deliberation of reality television's burgeoning popularity and compared various shows. The majority of journalists were astonished at the popularity of *American Idol,* as well as the entire genre of reality television.

In particular, many articles expressed surprise at the ratings achievable during the summer months. For the majority of broadcast television history, summer was a wasteland of reruns, devoid of new programming. With the increasing popularity of cable channels like HBO and Showtime, which introduced new series throughout the year, network television needed new shows to compete in the summer. The networks turned to reality television. Accordingly, the summer-debut rating successes of *Survivor* in 2000, *Fear Factor* in 2001, and subsequently *American Idol* in 2002 proved that audiences watched television in the summer. The press took notice. One reporter noted: "This new breed of cleverly conceived reality shows of summer . . . has displayed more originality and generated more excitement than virtually all of the far more expensive dramas and sitcoms begun in the last three years in the regular season" (Carter, 2002). Jeff Zucker, the president of NBC Entertainment, said, "It has to get your attention when three of the four biggest hits on television over the past three years have come out of the summer" (Carter, 2002).

Still, many journalists, particularly television critics, considered reality television to be terrible and lacking in any decent qualities. Much news coverage rhetorically questioned why reality television existed. Other writers knew a primary answer was cost. Reality television is cheap to produce and allows a way to experiment with new features, such as interactive voting and product placement. Previously, networks aired reruns of sitcoms and dramas as a way to amortize the cost of the original episode. For example, a repeat of *ER* "costs NBC nothing but pulls in a good chunk of ad revenue—thus helping to pay for the skein's roughly $7 million–$8 million license fee" (Adalian, 2002). However, with some channels providing new programming, the ratings for networks' reruns sank. Reality television proved to be an inexpensive means of regaining audiences.

Finally, news media noted that the incredible success of *American Idol*, just like the British version *Pop Idol*, was soon to be replicated elsewhere. Various articles reported that the format rights had been sold to producers in Poland, South Africa, Germany, France, the Netherlands, and Belgium, and that Scandinavian and Middle Eastern versions were planned.

American Dream

The rags-to-riches theme prominent on the show is also evident throughout the news coverage. In *USA Today*, Bill Keveney (2002a) explained the show thusly: "Take the American dream of stardom, mix in saucy English put-downs, and pop go the ratings." *Good Morning America* called the show "a Horatio Alger meets Britney Spears kind of story" (Brune and Ross, 2002b). Academics were also quoted about the success of *American Idol*: "'It's a rags-to-riches story,' said James Seguin, professor of communications at Robert Morris University near Pittsburgh. 'It's the success of the common man in a perfect television format.'" Seguin added, "People are living their dreams through these kids" (Gillin, 2002).

Likewise, *The New York Times* described the allure of the show as the "always-potent fantasy of being plucked from obscurity for stardom" (James, 2002). A *USA Today* trend article described the show's premise as the concept "that ordinary people can attain extraordinary recognition by virtue of ability and perseverance" (Gardner and Keveney, 2002). Gardner and Keveney (2002) also contended "the desire for instant fame and adulation that fuels such shows has never been stronger."

Clearly, reality television is not uniquely American. However, some American audience members may think that it is. One viewer noted, "'It allows persons from all backgrounds to utilize their talents,' says Alfonzo Webb Sr., 61, of Inglewood, Calif., who attended Wednesday's show with his wife and two adult children. 'That's the American way'" (Gardner and Keveney, 2002).

The press coverage also emphasized that the program *is* the contestants' dream. As Ryan Seacrest explained, "This is their dream. Every one of these kids has dreamt about being the next Britney, the next Marc Anthony" (Oldenburg, 2002). Justin Guarini agreed, telling Katie Couric on *Today* "I'm living my dream" (Wald, 2002a). He reiterated again on *Today*, this time to Matt Lauer, "I'm a small town kid, and, you know, this is my dream. And I'm—I'm living it and loving it" (Wald, 2002c). The contestants' dreams were not just of fame. Justin claimed:

> "I don't think the goal is to be a star. I think the goal is to do what we love. And what we love just happens to be something that can be very glamorous." Kelly Clarkson, 20, another finalist, concurs: "I want to be a performer. I want to have a job where I come home at night and I love what I do." (Gardner and Keveney, 2002)

Executive producer Nigel Lythgoe disagreed: "They would love to be stars. They would love to be recognized and applauded as singers, and with that comes stardom, whether you like it or not" (Gardner and Keveney, 2002).

Contestant R.J. Helton's mother was quoted in *The Washington Post* saying, "I always knew he'd be discovered . . . It was just a matter of being in the right place at the right time" (Waxman, 2002). Justin's mother agreed: "'It's surreal. We've felt like we've been taken to another planet,' she said breathlessly. 'Six weeks ago we were another family in Doylestown, Pennsylvania, mowing the lawn, living in our nice neighborhood. I was selling real estate'" (Waxman, 2002). Not only can the winner achieve her dream, but "the Fox network's summer sensation has generated enough fervent attention among young fans to launch a multitude of careers" (Hiltbrand, 2002). Thus, the 2002 news coverage echoed the rags-to-riches ideal present throughout the seasons of the program itself.

Overtones: Derision and Awe

Patterns, word choices, and phrases throughout the 2002 news coverage implied two general overtones. The first was strong disdain for *American Idol*. The second was awe and surprise at the number of people watching, the amount of money being made from the show, and the disregard of cultural standards. These two complementary connotations were often present in the same articles.

In *The Washington Post*, reporter Tom Shales derided *American Idol* as "roadkill TV—a gruesome spectacle that you can't help slowing down to stare at." He also labeled the cohosts as "fecklessly sappy" and called the show "more freak show than talent show." Moreover, he ridiculed the contestants, writing in one example that, "one of the least auspicious contestants is the fire-breathing Ryan Starr, who looks like she eats live mice for breakfast. She does have determination, though—determination befitting a fascist dictator." While Shales's exact point is unclear, he certainly is insulting. He ends with this:

> *American Idol* is indeed good for a giggle, but it's a cause for lament, too—style being passed off as substance again, attitude being substituted for ability, a perverse twisting of democratic ideals to encompass the notion that even the manifestly untalented deserve a spotlight, an audience and a cheering section of their own. (Shales, 2002)

In an example representative of many articles, Shales unremittingly criticizes the show, finding no redeeming qualities.

A similar article, titled "*American Idol* has the crass to make it must-see," appeared in the *Boston Globe* on September 4, 2002. In it, reporter Matthew Gilbert asked, "Why is it that dislikable things often make addictive television?" He thought that *American Idol* wasn't "some kind of altruistic attempt to give a young performer a chance at the big time; it's the first leg in a marketing campaign that will culminate in CD sales." While some writers asserted that the rags-to-riches narrative fueled the show, Gilbert contended it was driven by a "public-shame factor" and was truly an "interactive commercial." However, in an effort to explain its popularity, he observed the interactive voting component and an "ironic appeal"—it's so bad that it's good. However, neither these elements nor winking explanations could explain the immense popularity of the show.

Bashing *American Idol* occurred over and over. Karen Heller in *The Philadelphia Inquirer* called it a "junk show" and a "guilty pleasure," noting the "banter of hosts Ryan and Brian is evil. Recovering cheerleader Paula Abdul's hollow platitudes are evil" (Heller, 2002). Censure occurred on television as well as in print. Toure, an editor of *Rolling Stone* magazine, appeared on the *CBS Early Show* on September 5, 2002. He denounced *American Idol* at length, calling it "American Cheeseball." Toure also asserted that the "American people were a little bamboozled by this, you know. I mean, these people [the contestants] couldn't sing." Toure bemoaned the show's and the contestants' lack of hipness, alleging that people wouldn't buy *American Idol* associated products because they weren't cool enough. Of course, Toure also thought, "This is going to be like the Pet Rock and, you know, a year or two from now, you'll be like, Who won *American Idol*? Who shot J.R.? I don't remember." Almost a decade later, *American Idol* is still a top-rated program and has sold millions of ancillary products.

Several criticized *American Idol* while concurrently marveling at its success. For example, in *The St. Petersburg Times*, Eric Deggans (2002) dubbed *American Idol* a "glorified karaoke contest—a gussied-up, lip-synching carnival" and the "embodiment of what has gone wrong with network TV's summertime programming." However, the same article noted *American Idol's* stupendous ratings and that it was the "darling of the TV industry." Nonetheless, Deggans (2002) warned viewers that "Like a chocolate-covered ant, this TV show hides its awful qualities beneath a sticky-sweet coating." Perhaps the success of the show invited television critics to defame it. *American Idol* pandered to a large audience and is admittedly lowbrow, a formula that many critics apparently disliked.

M.S. Mason (2002) wrote in the *Christian Science Monitor*, "It's the summer's big hit. . . . But I have to confess that I found the first few weeks of *American Idol* intolerable." In *USA Today*, Robert Bianco (2002) declared that of the new reality shows *American Idol* is the "best and most popular of the group" but that at the same time it is "sadistic" and "cruel." Similarly, Caryn James (2002) stated in *The New York Times* that the show had "jaw-droppingly awful" moments, yet "the series does have a stroke of commercial genius, as it shrewdly combines elements from a smattering of other series into one big marketable soup." In *The Washington Post*, reporter Donna Britt (2002) asserted: "Just because *American Idol* was a summer smash . . . doesn't mean the TV show wasn't crass, shallow and cheaply manipulative. The series' production values were mediocre, its authenticity questionable, its hairstyling horrific." And yet the author looked forward to hearing winner Kelly Clarkson sing at the Lincoln Memorial 9/11 memorial concert, because of the American nature of her "winsome cocktail-waitress-turned-recording star" turn. Also in *The Washington Post*, Lisa de Moraes (2002) marveled at the ratings and commented on the competition winner singing at the 9/11 memorial, acerbically asserting that, "The terrorists have won."

Some journalists seemed to be shocked at the callous nature of the show, the humiliation of losing, and the program's low-culture nature that appealed to the largest audience possible. Nonetheless, instead of disgusting viewers, these components seemed to attract them.

2008 PRESS COVERAGE ANALYSIS

For the second discourse analysis, an initial Lexis Nexis search of the words *American Idol* between the dates of January 15, 2008, and May 21, 2008, returned 1,895 articles and transcripts. Pertinent stories about *American Idol* were found in diverse media, including *The New York Times*, *Advertising*

Age, *USA Today*, *New Musical Express*, and *The New Yorker*, as well as on network and cable television. Articles or transcripts that were primarily about programming notes, ratings, or recaps, or that only indirectly referred to *American Idol* were discarded, as were duplicates of articles. The Lexis Nexis search returned over 905 radio and television transcripts. Unfortunately, the majority of transcripts were of local radio shows that only mentioned *American Idol* in an offhand manner. Of the radio and television transcripts, only 45 were deemed relevant. However, the great majority of the 990 newspaper and magazine articles were included. Of the newspaper and magazine coverage, only 155 articles were discarded (15 percent), which corresponds to the exclusion rate of the 2002 study. Moreover, combining the numbers of relevant transcripts and articles yielded a total of 880 pieces of 2008 news coverage that were analyzed. The lower number of articles may be due to the seventh season's lack of newness or less coverage of television in general in 2008. This study found five themes in media coverage of the 2008 seventh season of *American Idol*. The themes included economics, authenticity versus artificiality, past contestants, current contestant stories, and specific contestant Carly Smithson.

Carly Smithson

Carly Smithson was an early favorite in the 2008 competition. However, her Irish nationality, prominent tattoos, and previous recording deal made her a controversial contestant. In particular, her earlier recording contract sparked dispute about her amateur status. An essential part of the Horatio Alger narrative is beginning as an unknown amateur. Could someone who already had a shot at stardom, a major label record release with millions of promotional dollars supporting it, be considered an amateur? In other words, does the Horatio Alger narrative include a second chance?

On *American Idol* a second chance is possible. According to the *American Idol* rules, Carly was allowed to compete despite that "under her maiden name, Hennessy, she released an album in 2001, *Ultimate High*, and, according to one report, MCA backed its promotion with $2 million" (Fink, 2008). Press also noted fan opposition to Carly's *American Idol* participation. Jill Dobson on the Fox News Network said some "fans online . . . say they'd rather see a rags-to-riches story than a contestant like Carly Smithson." It was a divisive issue. Some fans "prefer to see contestants that are straight from the fast-food counter to life in the fast lane, [but other] fans we talked to say it's more about the performances than the back story" (Snyder, 2008a). Other fans cried foul because Carly "is no amateur" (Wald, 2008). Meredith Viera and Matt Lauer agreed on the *Today* show that "amateurs should be amateurs" (Wald, 2008). Starting as an amateur is an integral, if unspoken, beginning of the rags-to-riches story.

The press implied that Carly, despite her vocal skill, seemingly lacked the necessary ingredients to fulfill the rags-to-riches story. However, Carly's first doomed shot at fame was molded into part of her hard-luck tale. She immigrated to America for her chance at stardom, but failed miserably. This failure was her hard-luck story, now able to be redeemed through *American Idol.*

Carly Smithson received her second chance at achieving the rags-to-riches narrative, as the press noted. She triumphed by being able to compete on the show despite her initial ill-fated attempt at fame, a decidedly not-mainstream appearance due to the tattoos, and her foreignness. She subsequently finished in sixth place. *American Idol* arbitrated the changing nature of the American rags-to-riches story. People can receive a second chance to accomplish fame and fortune—at least on television, if not in the press.

Contestant Stories

As in 2002, in 2008 stories about the various current contestants were a human-interest element of the press coverage discourse. These articles provided a few details of contestants' lives, and during the auditions, of potential contestants. The initial articles concerning the auditions often focused on contestants' emotional stories, while articles about the top-twelve contestants often repeated known facts. This reiteration of facts reduced the contestants from multifaceted, complex people into familiar archetypes.

Hard-luck stories comprising the beginning of the rags-to-riches journey were plentiful. For example, auditioner Angela Martin's story was mentioned in numerous press outlets. Her daughter was afflicted with Rett syndrome, a neurodevelopmental disorder (Hoffman, 2008a). Other auditioners' human-interest stories included a man who lost 205 pounds, someone who collected fingernails, and an overweight woman trying to achieve her dream.

In general, though, news filled in knowledge gaps. Press coverage explained contestants' backgrounds to those who may have missed them on the show, or to those who desired more information about the contestants. However, without interviews with the contestants (in 2008 reporters had no access to current contestants), the majority of articles merely repeated certain key information: hometowns, ages, jobs, marital status, sad stories (for example, David Cook's cancer-stricken brother or David Archuleta's overbearing father), musical genre preference, and one or two fun facts (such as Kristy Lee Cook liking horses). This may have piqued interest in news consumers, provoking more people to watch the show.

More importantly, the rote repetition of facts diminished each contestant to an archetype. Being unaware of many specific personality traits and life experiences allowed the home audience to fill in the contestants' life stories with speculations and ideas gleaned from previous contestants or other mediated experiences. In a sense, the contestants were blank slates, able to be shaped into what the audience could easily understand. Beyond certain undeniable facts, each contestant could be relegated to a certain archetype, either one already present on the show in past seasons, or one familiar to the audience from past media consumption.

Where are They Now?

Another contestant-based theme also existed—the "where are they now" update. In general, two discursive frames existed. The first was negative, for those contestants who did not succeed in maintaining a modicum of fame after appearing on *American Idol*. The second was positive, for those contestants currently working in the entertainment industry. Unsuccessful former contestants were linked with blue-collar jobs, a return to the beginning of their rags-to-riches narrative. Successful previous contestants were associated with fame and illustrated as hardworking. Their work ethic was framed as having helped them gain success. The converse of this is the inference that lack of success may have been caused by laziness. In addition, past male contestants were framed in terms of economic and celebrity success (especially if not achieved) while past female contestants were also construed as successful in other terms, such as education or homemaking.

The primary discursive frame associated with previous contestants was negative. This "where are they now" discourse implied that success was not achieved and that the former contestants were living in ignominy. On CNN's *Showbiz Tonight*, host Brooke Anderson asked, "Why are the winners and the runners-up having so many problems after they are off the show? Why are they getting dropped by their record labels?" (Levine, 2008). While noting that many winners do achieve success, *Showbiz Tonight* could "reveal that winning is not a guarantee of great success." The program cited as evidence the fact that executives at record labels have dropped many former contestants, including past winners Ruben Studdard, Taylor Hicks, and runner-up Katherine McPhee. Of course, their seeming failures may be countered by the success of Kelly Clarkson, Chris Daughtry, and Carrie Underwood. However, as reporter Ben Tracy noted on *The CBS Evening News* "other *Idol* winners' music careers have become, well, idle" (Kaplan, 2008). This discourse relied heavily on negative framing.

Male contestants who did not achieve lasting stardom were portrayed as unfulfilled and still hopeful of achieving glory. In contrast, the female contestants who did not achieve popularity were usually depicted as happy to be homemakers or students, content with their brief period of fame while on the show. *USA Today* featured several columns that caught up with former participants, including season-six contestant Stephanie Edwards. An article featuring her noted "While most of her fellow finalists are chasing stardom, she's chasing a degree at Armstrong Atlantic State University in her hometown of Savannah, GA" (McGinn, 2008b). Stephanie was depicted as prudent and conscientious for focusing on her education and not "putting all her eggs in one basket" in an attempt to attain fame. In a marked difference, season-two contestant Rickey Smith was profiled in the same column a couple weeks later. Rickey had tried unsuccessfully to land a record deal in Los Angeles after the show. The reporter descriptively claimed that "like a twister on the Plains, reality tore the corrugated roof right off his plans" (McGinn, 2008c). Rickey, having depleted his savings, was forced to return to Oklahoma City: "Says Smith, who turns 29 Saturday 'I wasn't ready to accept the fact that I had to come back and give up music. I've made peace with it, but it's still frustrating'" (McGinn, 2008c). While working as a bartender ("a necessary evil" according to the writer), Rickey recorded an unreleased independent album and still planned to return to L.A. to search for a record deal. Rickey liked Oklahoma, though: "'Anywhere I go,' Smith says, 'they tell me I'm still a celebrity. I appreciate that'" (McGinn, 2008c). Whereas Stephanie Edwards was lauded for her decision to finish her college degree, Rickey Smith was painted as a loser, unable to achieve stardom or a middle class job.

However, the success discourse was a striking contrast to the failure discourse. This discourse included stories of successful contestants working in the entertainment industry, either recording albums, touring, or singing on Broadway. Broadway and off-Broadway performers included Tamyra Gray, Ruben Studdard, Clay Aiken, Diana DeGarmo, LaToya London, Fantasia, Anthony Federov, Lakisha Jones, Jennifer Hudson, Taylor Hicks, and Constantine Maroulis. In an example, Constantine Maroulis released an independent album, toured in support of it, and appeared on Broadway and *The Bold and the Beautiful* (McLaughlin, 2008). In 2009, Constantine continued his success, starring on Broadway in "Rock of Ages" and earning a Tony Award nomination. In another example of success discourse, season-two contestant Carmen Rasmussen was also noted as hardworking, with a new album, a book, and a potential reality program based on her life (McGinn, 2008a). Fourth-season contestant Jessica Sierra achieved an infamous kind of fame, if not success—numerous arrests and then participation on *Celebrity Rehab with Dr. Drew*, which aired in the spring of 2008.

Authenticity Versus Artificiality

Throughout the 2008 press coverage, reporters debated the authenticity and artificiality of the show. The premise of *American Idol* is an earnest and realistic depiction of a search for a star. News discourse questioned the veracity and sincerity of the show itself. Was the show as truthful and genuine as it alleged? Or perhaps was the show already planned and scripted? Moreover, while *American Idol* is an undisguised search for marketable talent, many journalists decried the stereotypical physical appearance they assumed the show endorsed, particularly being thin and attractive.

As noted previously, *American Idol* endorses already existent beauty stereotypes. The press coverage observed this as well, focusing on the importance of physical attractiveness. One article, titled "Budding star aims to break the *American Idol* mold," described the attempts of plus-size singer Joanne Borgella's attempts to build an entertainment career (Schapiro, 2008). In another example, Gina Glockson, former contestant and cohost of *American Idol Extra* on the Fox Reality channel, discussed the 2008 contestants on *The Big Story with John Gibson*. She believed several of the contestants were talented, saying, "This year, I am so impressed with how talented the singers are, and how it's not really about how someone's hair looks or how someone's legs are beautiful. It's actually about singing talent" (Snyder, 2008b). In other words, sometimes the show *is* about body parts. While Gina insisted this year's competition was primarily about talent, a number of media asserted that appearances usually matter. For instance, some articles focused on the contestants' fashion choices. One such article claimed that "even though *Idol* is a singing competition, like any contest—from beauty pageants to job interviews—we all know that points are awarded, or deducted, for style" (Rodman, 2008). Another article in *The New York Times* commented on the predictability of the show, mentioning certain repetitive contestant archetypes (Wyatt, 2008a). One newspaper article, titled "No one can be authentic on a show this rigid," noted that "there are no hefty, ugly, bald or weird contestants this year" and that "*Idol* gets closer, every year, to manufacturing stars the way stars like Justin Timberlake were manufactured" (Crosbie, 2008).

One reporter declared that,

> to watch this show and think it is randomly prizing talent from thin air, and letting that same talent blossom, authentically, is to be a fool who doesn't understand the way the show works, or the way the recording industry works, for that matter. (Crosbie, 2008)

In *The New York Times*, Edward Wyatt wrote: "From the placement of local sorority members along camera sight lines to the instructions to the audience members about when to stand and how to wave their hands, *American Idol* is as scripted as a 'reality' show dare be" (2008). Nevertheless, while the premise of the show relies on the transparent process of manufacturing an audience-chosen star, there remains within the show an air of honesty, sincerity, and earnestness in this quest to find an undiscovered talent. Paula Abdul brought this sincerity into question in 2008.

On the April 29, 2008, performance show, Paula Abdul made a very public mistake. She attempted to critique two of contestant Jason Castro's performances when he had, in fact, only sung once. Paula explained her gaffe by saying she confused Jason's dress rehearsal with the show itself. This called into question the integrity of the show and prompted viewers to speculate about the unscripted nature of the judges' comments. Numerous articles and TV shows questioned the truthfulness of the show, as well as Paula's sanity and/or sobriety. This debacle, while giving the show more publicity, also incited doubt about the "reality" of the show, or its sincerity. Cecile Frot-Coutaz, chief executive of FremantleMedia North America and an executive producer of *American Idol,* noted in *The New York Times* that "judges are never given scripted comments and that she does not believe that the credibility of the show has been compromised" (Wyatt, 2008d). This line of thought was promoted in many articles. The show's honest and earnest nature in its mission to find a superstar remained untainted. Paula received the blame for her blunder.

Economics

Just as in 2002, *American Idol* economics was a dominant theme throughout news-coverage discourse. Within this coverage were various discursive emphases, including a focus on the amount of money the show earns, reports on new synergistic agreements between the show and sponsors, the effect of the show on the music industry, the charity works of *Idol Gives Back*, and various countries' indigenous versions of *American Idol.*

In a parallel to 2002, product placement was featured heavily in the economic press coverage. In 2006, *American Idol* featured 3,174 product placements. By 2007, *American Idol* had the most product placements of any show on television, with some 4,348 occurrences of product placement (Tracking Advertising Dollars, 2008). While in 2008 *American Idol* had a small dip in ratings (although it still was the highest rated program on television) the four core sponsors "have helped to make the business 'stronger than it has ever been,' says Robert F.X. Sillerman, chief executive of CKX, the firm that co-owns the franchise with FremantleMedia" (Lieberman, 2008). In the fourth quarter of 2008, CKX reported an *American Idol* generated revenue of $28.8 million, up 42 percent from 2007, explained by more hours of *American Idol* on television than the previous season (Lieberman, 2008). In *The New Yorker*, esteemed music critic Sasha Frere-Jones (2008) explained that the vertical integration of *Idol* "makes money for [creator Simon] Fuller while it is finding him pop stars who then go out and make him more money when they work as pop stars." No matter the ideals or archetypes *American Idol* propagates, in this revenue-driven economic theme the show is rightly depicted primarily as a corporation.

In this sense, *American Idol* was an equal to and able to partner with other large corporations. In 2008, Apple became a sponsor of *American Idol*. The press reported that iTunes had exclusive rights to contestants' performances, selling them as individual audio and video downloads. iPods became *Idol's* official digital music players, while the iPhone became the official phone (Schneider, 2008). ITunes also sponsored the online streaming of contestant performances on the official *American Idol* website, "where it also will have a substantial branded presence" (Yourse, 2008). Similarly, for a second year in a row, fast-food chain McDonald's also joined forces with *American Idol*, producing an *American Idol* Happy Meal, which was advertised on television, radio, and in print. Online, fans of McDonald's and *American Idol* could "vote for their favorite Happy Meal toys and enter a sweepstakes for a trip to the season finale of *American Idol*" (Maul, 2008). These synergistic relationships were mutually beneficial for the linked companies. Furthermore, McDonald's and iTunes provided more media presence of *American Idol*, contributing to the show's dissemination into American culture.

American Idol's immense popularity also influenced the music industry. As Paula Abdul said, "I think that, with the success of the show, people are realizing that it's a force to be reckoned with in the music industry. There is a real payoff dividend. It can happen" (Kinon, 2008). Taking her own advice, Paula jumped on the *American Idol* musical money train as well, releasing a single on a Randy Jackson-produced album titled *Randy Jackson's Music Club*. She and Randy did the rounds of press, promoting her single and the album. Other articles noted influence on current charts. Alison Stewart, reporting on National Public Radio's *The Bryant Park Project*, described the "*American Idol* Effect." She explained that contestants "sing classic songs during the TV talent competition and the next morning, songs by the likes of Stevie Wonder or Peter Frampton shoot to the top of the iTunes most-downloaded chart, displacing, let's say Fergie or Chris Brown" (Hoffman, 2008b). Bon Jovi, Aerosmith, and other artists received noticeable sales bumps. In an example, after contestant Jason Castro performed "Hallelujah," the 1994 Jeff Buckley song received the "largest song spike we've ever seen from an *American Idol* performance" selling 178,000 downloads (Mayfield, 2008). Furthermore, Jason Castro's performance of Hawaiian musician Bruddah Iz's version of "Somewhere Over the Rainbow" caused a 700 percent increase in sales, pushing the 1993 song from number 187 to number 11 on the Nielsen SoundScan's digital songs chart (della Cava, 2008). Interestingly, while the contestants' performances sell as iTunes downloads as well, they are not allowed to chart. According to iTunes, "iTunes and Fox are committed to presenting contestants in a fair and balanced manner online and on-air . . . for this reason, sales performances . . . will not be reflected in the iTunes charts" (*American Idol* puts its tunes on iTunes, February 20, 2008).

Idol Gives Back, a special show that raised money for charities, was also particularly noted in the coverage. Press reported on which celebrities would be appearing in 2008, including presidential candidates Hillary Clinton, John McCain, and Barack Obama. Journalists also recounted that the money raised in 2007, an estimated $76 million, was distributed to at least nine charities over two years through the Charity Projects Entertainment Fund (Wyatt, 2008b). This fund is the infrastructure through which Comic Relief, another entertainment-based charity organization, raises and disburses its donations. In 2008, a newly created charity organization, *Idol Gives Back,* distributed the money raised. This allowed *American Idol* to focus more on its own charitable causes (Wyatt, 2008b). In general, the reporting concerning *Idol Gives Back* was positive and a promotional boon for the show.

For example, Randy Jackson spoke on *Larry King Live* about the good works of *Idol Gives Back*, noting that the money raised educated "kids learning to read and also about healthy lifestyles." Ryan Seacrest, Simon Cowell, and Paula Abdul, also appearing on *Larry King Live*, explained that *Idol Gives Back* wasn't simply about raising money, but was about entertaining audiences as well. Through entertainment, *Idol Gives Back* raised awareness without "preach[ing] at the audience." As Ryan explained, "we are trying to use that vehicle, that machine that is *American Idol* to create a little bit of awareness when it comes to problems here in our own backyard with kids and also kids around the world" (Douthit and Whitworth, 2008). The show's noblesse oblige masked the relatively little amount of money raised. In contrast to the amount of money the show itself earns annually, not to mention the enormous sum raised over seven years of production by 2008, *Idol Gives Back* donated a paltry figure. According to *Forbes* magazine, Simon Cowell's 2007 *American Idol* salary was $45 million (The Celebrity 100, 2008). Moreover, *Advertising Age* valued *American Idol* as worth at least $2.5 billion in 2007 (Atkinson, 2007). In 2007, *Idol Gives Back* raised $76 million (Wyatt, 2008b). The amount donated by the show was relatively little in comparison to the amount it earned.

Additionally, by 2008, *Idol* franchises were springing up around the globe. As one article noted, "*American Idol* is broadcast to over 100 nations outside the US. Aside from the Philippines, the show is aired live in Pakistan, India, Israel, Canada, Indonesia, Malaysia and the United Arab Emirates" (Valisno, 2008). Moreover, while the American version airs globally, indigenous versions also exist in several countries. Other *Idol* franchises include the Brazilian *Idolos* and the pan-regional *Latin American Idol*, which holds auditions in Mexico, Argentina, Guatemala, Colombia, and Venezuela (de la Fuente, 2008). Many non-American versions received notice in the American press, particularly the Afghan and Chinese versions. Within these cultures, western components of the show violated traditional cultural norms.

On *Afghan Star*, a female contestant flouted the Taliban traditions banning both music and women from public performances. According to *Good Morning America*, 18-year-old Lima Sahar was "on her way to becoming a pop queen and a symbol of freedom and courage" (Brune, Ross, and Sherwood, 2008). Sahar was the first woman ever to reach the final three, and *Good Morning America* called her success "a triumph over the Taliban." While she sang conservative folk songs, wore a head scarf, and didn't dance, her very presence on the show indicated progress for women. When faced with criticism Sahar avowed "that there can be no progress made for women without upsetting the status quo" (Brune, Ross, and Sherwood, 2008). Similarly, the Chinese version, *Super Girls Singing Contest* (which featured only women), also illustrated potential social change. *Super Girls Singing Contest* aired nationally in China in 2004, 2005, 2006, and 2009. The Chinese version of *Idol* was intensely popular: 400 million people watched the 2005 season (Hoffman, 2008c). Jian Li, director of a documentary film titled *Super, Girls!* about the Chinese competition, asserted on NPR that the *Super Girls Singing Contest* was a first taste of democracy for the Chinese people. He emphasized that voting for the first time, even for a singing competition, provoked a new set of emotions: "You feel you can make a difference. You know, you feel like never before that you can make a difference in somebody else's life, and your vote counts. That's how you felt" (Hoffman, 2008c). The western ideals present in *Idol*, gender equality and democratic voting, may instigate social change in China and Afghanistan, or at least make these ideals visible in a national format. Ironically, while the western ideals present in versions of *Idol* may challenge other countries' traditions, in *American Idol* the western ideals support the traditional status quo.

COMPARISON OF 2002 AND 2008 PRESS COVERAGE

The 2002 and 2008 press discourses had many similarities. Both contained economic themes, a focus on contestants, and a discussion of the Horatio Alger theme (within the contestants' stories in 2008). The 2002 and 2008 years' economic press coverage featured both the portrayal of the program as a business and an emphasis on product placement. Each year's press coverage also noted the importance of the rags-to-riches narrative within *American Idol*. The 2002 coverage noted its presence in the premise of the show, while the 2008 coverage interrogated who was entitled to this journey and if the show was honest in its depiction. Additionally, the failure discourse in the 2008 "where are they now" theme questioned if the show could provide the end of this narrative for all contestants. The 2002 coverage introduced news consumers to the contestants and distilled each contestant to one particular wholesome trait. The 2008 coverage continued this trend with one difference: it assumed consumers were familiar with the *Idol* archetypes, and by presenting a limited number of facts, allowed consumers to draw their own archetypal conclusions.

The two insinuated themes of the 2002 coverage, awe and distain, were not present in 2008. While there was some negative coverage of the show, mostly lambasting it for predictability and dullness, in no way was this comparable to the viciousness of the 2002 denigration theme. By 2008, *American Idol* was an established show, the leader of television ratings for most of a decade, and producer of some credible musicians. Overt denigration and dislike were hardly present in the 2008 coverage. Similarly, the awe theme was also absent. The success of *American Idol* in 2008 was no longer shocking, but to be expected after the several years of ratings dominance. While there were many similarities between the two discourse analyses, it is worth noting that the nature of the press coverage changed, as indicated by the 2008 absence of the overtones present in 2002.

Within the overall press coverage, there are several types of media and media coverage present. The types of media and coverage range, for example, from human-interest stories on the radio to newspaper and television criticism to publicity-generating interviews on *Larry King Live*. Commentary and reviews are inherently critical, while other pieces seem to directly refer to press releases and perform public relations duty. As *American Idol* gained larger audiences, it's possible that the amount of publicity for the show grew as well. This may account for the larger number of public relations stories present in 2008 than in 2002. In 2008 there was also a significantly smaller amount of critical commentary directed at the show. One reason for this may be that television critics in 2008 were less likely to criticize a program proven to be immensely popular. As more people seemed to enjoy the program, commentary belittling it may have seemed to insult viewers as well. An additional cause for less commentary in 2008 may have been the shrinking of the newspaper industry. By 2008, fewer reporters were devoted solely to television commentary. Perhaps those reporters covering several media outlets were less likely to write about an established program such as *American Idol.*

In sum, a primary focus on economics, finances, and power remained the same between 2002 and 2008. Concurrently, contempt for the show disappeared. Both of these findings may be explained, at least in part, by the increased popularity of the show. As the show grew in popularity, so did the subsequent amount of revenue. The program could charge more for advertising, and it increased the number of revenue streams. *American Idol* is a paragon of vertical and horizontal media integration, with enormous potential for earnings. Simply put, *American Idol* is a ratings and profit monster, earning massive amounts of money. This could hardly go unnoticed. Similarly, the massive popularity may have inhibited many critics from commenting negatively. To insult the show is similar to insulting consumers of it, many of whom also presumably consume news.

However, the popularity of the program accomplished more than simply earnings. By 2008, *American Idol* had fully entered American mainstream society. It was enmeshed into American culture through the many media facets in which it appeared. Contestants and the judges appear throughout contemporary life in media including magazines, newspapers, online venues, and music. They win Academy Awards, appear on television, top the charts in various musical genres, and tour around the country in musical productions. *American Idol* has been a significant part of modern American life since 2002. The show contains several American ideals, ranging from the rags-to-riches narrative to current beauty and fashion paradigms. *American Idol* represents American principles and the existing status quo. Marvel, awe, and simple reporting about *American Idol* economics are permissible, but derision for the show is apparently unacceptable. At this point, contempt for *American Idol* may be almost tantamount to contempt for America and American ideals.

Chapter 10

Conclusion

As each season of *American Idol* drew to an end, the number of contestants dwindled. As each contestant left the show, the program highlighted his or her *American Idol* experience with an elimination video recapping his or her *American Idol* journey. The highlight reels consisted of edited-together clips of phrases the contestants had said while on the show. These videos, particularly the contestants' statements, revealed that the contestants themselves often internalized several of the ideas present in this book. These highlight videos also provided an end to each contestant's journey, completing his or her narrative on the show. An examination of the elimination highlight videos permits an illuminating summary of the ideals and themes found in this book.

Season-three contestant Camile Velasco was voted off the show on April 7, 2004. Camile's highlight video voice-over said:

> I'm so glad I decided to audition for *American Idol.* Couple months ago I was waiting on tables, now I'm singing songs for 30 million people. It's amazing. This whole experience has definitely been very emotional for me. There were times during the audition process where I thought I wasn't going to make it. I cried and I prayed and when I did make it, I cried again. It's just been really overwhelming. The thing that I've learned with this whole experience is to believe in myself more. And that's really what I need to do, use criticism to be a better performer. I'm so thankful for this experience. Mahalo. (April 7, 2004)

Camile spoke about her time on the show, arousing demonstrable emotions, which helped illustrate her personality. Moreover, the word *Mahalo* was a nod to the people within the Hawaiian regional collective identity and a symbol of her allegiance and belonging to that group. At the same time, affiliation with Hawaii identified her as American, completing her immigrant story. She had become a famous American.

Third-season contestant Amy Adams discussed different ideals. She said:

> It's really wild how your life can change in a week and then everyone knows who you are. It's so intense and so instant all at the same time. The most exciting thing is being in front of the live audience, where you actually get to see, okay, this is who we're doing this for, this is actually the beginning of our journey. I've been granted a gift from *American Idol* to kinda get my voice out there and share who I am with America. It's been an amazing, amazing thing, an amazing ride. (March 31, 2004)

Amy Adams was shocked at how quickly her celebrity occurred. The live audience of adoring fans helped transform her from an anonymous person into a celebrity. Simultaneously, she was able to inform America of her archetypal identity. Her girl-next-door archetype, in conjunction with the adoring audience (as well as other factors), helped her become a celebrity.

Jennifer Hudson also noted how she was able to succeed on *American Idol*. She explained her emotions:

> When I got the call for wildcard, I was overjoyed. Oh god, so grateful, so excited. To know I get to come back another time, get a third chance, when this is a one-chance type of biz, you know. I had a lot of emotions going on. I am a very out-spoken type of person. I feel what I think. You'll see it through my clothes, you'll see it through my personality, you'll see it. It just comes out and there's nothin' I can do about it. This show's about versatility. It's been a struggle. It's hasn't been easy, but if it's not worth working hard for, it's not worth it at all. So, I'm proud of the struggle that I've been going through. And I'm making it; I'm a survivor. (April 21, 2004)

In order to achieve celebrity, Jennifer knew personality needed to be visible even through her clothing choice. Moreover, her tears and emotions added to her supposed emotional authenticity. Jennifer also referenced the hard work required for the *American Idol* rags-to-riches journey. Her rags-to-riches journey was necessarily difficult and provided the requisite fame-and-fortune ending.

Finally, season-seven contestant Chikizie brought up another important *American Idol* ideal. He emphasized the backstage nature of the program by discussing a commercial shoot and his family. Chikizie said:

> Succeeding in the show would be a really great chance to give back to my parents because they've done so much for me. I'm just really proud to be here; I feel really blessed. At a commercial shoot. It's really insane. I mean, now people

know my name. At rehearsal, photo shoot, onstage. It's been an amazing ride. I hope it lasts forever. (March 26, 2008)

As Chikizie stressed, family was a recurring theme throughout all seven seasons of the show. Moreover, it was a recurring theme through all the aspects of the show, emphasized in many archetypes and the blue-collar background typical of the rags-to-riches journey. Family was a uniting link for everyone on the show and in the audience, despite demographic differences. Everyone on the program professed to value their families, no matter their race, class, or place of origin. While sibling relationships were mentioned, the primary relationship praised on the program was the parent-child relationship. Not only were contestants' relationships with their parents stressed, contestants who had children frequently cited them as a motivating factor to succeed. In season one, Nikki McKibbin was defined as a single mother. Season-two contestant Josh Gracin also had children. On season three, LaToya London's two stepchildren were duly noted. Moreover, Fantasia's toddler frequently appeared in the actual theater where the show filmed. Season-four contestant Scott Savol had a son who was not mentioned often, presumably because the mother had custody. On the other hand, season-five contestant Chris Daughtry had two stepchildren who were positioned as his own children. In season six, contestants Phil Stacy and Lakisha Jones both had children. Season seven was the only season when none of the top twelve contestants had children. By contrast, two season-eight contestants, Megan Joy and Alexis Grace, were single mothers. Season-eight contestant Lil Rounds also had three children.

Parents and other family members were often shown reacting in the audience as their relatives sang. More importantly, however, mothers and fathers of contestants were often interviewed about their children on the show and in the press. Each set of parents spoke about their child's natural talent, reinforcing authenticity, and about what he or she was like as a child and the bond they shared. While the standard nuclear family was not put forth as a model (families of almost all compositions were shown, excluding same-sex parents), family itself was a clear ideal.

AMERICAN THEMES

The first research questions posed are about contestant narratives in *American Idol*. What contestant narratives are present in *American Idol*? Within these narratives, what do ideals and discourses imply about America? Each chapter reveals aspects of American culture through the lens of *American Idol*. As

Fiske (2000, p. 220) noted, television often "make[s] meanings that serve the dominant interests in society." So did *American Idol.*

Through the overarching narrative of the search for a superstar, contestant narratives, and show discourse, *American Idol* clearly endorsed the rags-to-riches narrative, contemporary archetypes, and a celebrity ideal. It did this in a seemingly transparent manner, inviting the audience to participate in the creation of these ideals. However, the audience was still susceptible to creating celebrities that conformed to typical standards. Thus, cultural standards remained in place.

In addition, *American Idol* supported the existence of the discourse of dreams and the rags-to-riches narrative journey. The show promised wealth and fame to the victor who necessarily established his or her own rags-to-riches journey. This journey was emphasized through hard-luck stories, blue-collar backgrounds, and immigration tales. The difficulties explained in the contestants' narratives established that the contestants deserved the ending of the rags-to-riches journey. The show discourse stressed that, through the show and its rags-to-riches promise, dreams could come true.

Other American ideals manifested throughout the show discourse were archetypes. Each archetype represented a segment of the American population. As discussed in chapter four, *American Idol* contestants typically fit archetypal categories. These archetypes were models that helped audiences make sense of the world (Lule, 2001). These archetypes, while present on the show, were originally created in other media. Alternately, *American Idol* also adapted some archetypes into new versions.

For example, the rocker archetype contained reliable elements, such as a certain sartorial appearance and preference for rock music. Nevertheless, the contestants embodying this archetype varied over the years, including a rocker dad and an emotional rocker. This archetype was highly visible in season eight, with a new twist. Second-place finisher Adam Lambert was clearly a member of the rock archetype, as displayed through his dyed black hair, skintight jeans, and preference for rock and roll. However, he bent gender demarcations and showed a predilection for a genre referred to as glam rock, exemplified by such famous musicians as 1970s era David Bowie and the rock band Kiss. His performances were almost high camp, and included heavy make-up, platform boots, and scale-defying vocal stretches. Adam's appearance and performance style, as well as his subsequent confirmation of his homosexuality after the season ended, added a new twist to the rocker archetype. At the same time, he was undeniably season eight's resident rocker. Similarly, Kris Allen, the winner of season eight, was a clear continuation of the singer-songwriter and the guy-next-door archetype. Kris strummed the guitar, looked cute, and earnestly sang songs that appeared as if he had written them, just as Jason Castro had done the previous season. Presumably both Kris's and Adam's clear archetypes aided their popularity. Each fit an identifiable archetype and thus was easily understood and familiar.

Interestingly, *American Idol* archetypes omitted any discussion or mention of race, sexuality, or gender issues. For example, no one in the rhythm-and-blues archetypal category ever discussed what it was like to grow up as an African American. Even as photographs surfaced of contestant Adam Lambert kissing another man, he refused to comment on his sexuality. While his reticence may have been explained by several different reasons, he (and the show discourse) was silent about his sexual preference. This silence contributed to a diminishment of public dialogue about sexual preference issues. Discourse about race, gender, and, to an extent, class, was similarly missing from the show. Discussing issues such as these would have broadened the audience's awareness and knowledge. In effect, silence erased these issues. Silence also reinforced the race-blind rags-to-riches success possibility and ignored the fact that not everyone has equal opportunities.

The archetypes also signaled several American racial and sexuality norms. African Americans were restricted to the rhythm-and-blues archetype, while white contestants appeared in nearly every archetype. Mixed-race contestants and those with Latino ancestry were frequently positioned as the benchmark, Caucasian, through a lack of discourse about their heritage. Asian and Indian contestants typically appeared in the girl- or guy-next-door categories. This confinement of certain races to certain archetypes limited how contestants were viewed and understood. The show supported American norms, such as one that restricts African Americans to rhythm-and-blues archetypes. There was not one African American country or rock singer, male or female. Alternately, the presence of Caucasian people in every archetype (except notably the female rhythm-and-blues diva) connoted that there are few limits for Caucasian people. White men were even in the urban and soul archetypes. Caucasians seemingly may inhabit whatever societal roles they desire. Asian contestants (all female) were not as limited as the African American contestants, yet still appeared only in two archetypes, the girl next door and urban singers. Additionally, all contestants were frequently depicted as sexual, except noticeably a portion of the African American men, who were often considered heavy or not conventionally attractive. In contrast, overweight African American women typically were portrayed as very sexual. Furthermore, sexual contestants were all positioned through discourse as heterosexual, confirming the hetero norm usual in American society and media.

This research demonstrates that *American Idol* clearly supported several American norms. This complicates Waisbord's (2007) contention that reality-TV formats embody neutral values. The format may carry inherent

Western values, but allows for national cultural values to be present as well. Western cultural standards were inherently present throughout the *Idol* format, such as support of democracy through the voting design, support of celebrity through veneration, and support of individualism through the ideal of the rags-to-riches individual success. For example, the Afghanistan version of *Idol*, *Afghan Star,* which employed a format similar to *American Idol*, supported Western ideals that are contrary to indigenous culture; the aforementioned democratic ideal, equality between men and women, who may both compete, and others.

While the show format was not value neutral, it did allow for some cultural standards to vary contingent upon the country of production. As such, the format of *Afghan Star* contained national ideals about beauty and also Western ideals of democratic voting, among others. This was apparent in a recent documentary—also titled *Afghan Star* (2009)—about the Afghan version. In this documentary, a female *Afghan Star* contestant waxed rhapsodic about her ideal man: "he should be of medium height, have strong eyebrows that meet in the middle, and I will love him" (Marking, 2009). Like *Afghan Star*, *American Idol* contained Western ideals inherently in the format of the show and also some uniquely national ideals through the discourse.

CELEBRITY

This book also answers the second research question: In *American Idol*, how is celebrity created and represented? Celebrity was created on *American Idol* through the narratives of status confirmation and authenticity. Authenticity and confirmed rank combined in *American Idol,* creating one overarching celebrity narrative. The ideal contestant had a great voice, was emotionally demonstrative, fit into an established archetype, and yet also brought something new to the competition. This emphasis on celebrity in the show naturalized the emphasis on it within American culture.

Chapter five explains how contestants are depicted as authentic and thus worthy of adoration. An inherent part of the reality-television fame narrative is that anyone can appear on reality television and achieve fame through uninhibited exhibition. However, *American Idol* also required talent in order to succeed, modifying the typical reality-television claim to fame. The *American Idol* contestants had to prove their authenticity through talent and realistic emotional authenticity. Concurrently, each contestant was required to adhere to who he or she claimed to be, or in other words, his or her archetype. Perhaps the most significant aspect of authenticity was what Hill (2002, p. 337) called "moments of authenticity" or the display of true emotions. Audiences expect reality-TV contestants to act (Hill, 2002). This falseness is countered through the moments of authenticity, or the display of true emotions. On *American Idol*, the behind-the-scenes perspective allowed the presentation of emotions, ordinariness, and archetypal characteristics.

Next, chapter six explains how *American Idol* confirmed contestants' celebrity status. On the show, celebrity was justified through verbal discussion of the contestants as celebrities, association with current celebrities, positioning of the contestants as celebrities through photo shoots and red carpet events, and image changes, as well as fan adulation. Moreover, the audience was complicit with the celebrity creation. They participated by being fans and helping to choose who would be the celebrities. Furthermore, the judges' comments suggested who was appropriate to be the American idol. As Marshall (1997) asserted, celebrity images may represent dominant social codes. Indeed, *American Idol* heavily relied upon dominant social codes for celebrity confirmation, including standard beauty and sartorial ideals.

Thus, emotional demonstrations, archetypal resonance, and talent created a seemingly authentic celebrity while discourse positioning confirmed this celebrity. The contestants were extraordinary through talent, yet at the same time ordinary through their emotions. As noted previously, celebrity promotes existing institutions, ideologies, and social systems (Marshall, 1997; Rojek, 2001). Additionally, celebrities are expressions of social values and role models (Evans, 2005a). The *American Idol* celebrities, constructed and authentic, also particularly embodied the culturally American values of beauty, hard work, and talent. The ideal of beauty promoted is typically a natural beauty, enhanced by celebrity accoutrements. The majority of contestants were attractive, thin, and wholesome. Each professed to work hard to refine his or her natural vocal skill and each immensely valued family, particularly children and parents. The majority of contestants emphasized a blue-collar and working-class background and possessed the typical rags-to-riches narrative. As celebrities, each contestant represented American ideals.

COLLECTIVE IDENTITIES

Chapters seven and eight address these research questions: How may *American Idol* illustrate American collective identity? What do the online discussants' conversations and contestant narratives connote about possible collective identity of an *American Idol* audience? Chapter seven details how show discourse, particularly direct-viewer address, formed a collective identity informed by American values and ideals. Chapter eight notes that participating in online discussion boards also helped form collective identities related to the show. In particular, two types of collective identities formed online: smaller identities formed around particular contestants, while an umbrella collective identity also existed as fans of the show itself.

The show discourse identified certain collective identities. These included an overall collective identity of the audience as American, contestant fan-group collective identities, and regional collective identities. The national American identity created by the show was an ideal, free from off-screen identity complications. This corresponds with Darling-Wolf's (2011) research on the French program *Star Academy*, which has a format similar to *American Idol*. Her textual analysis of the 2007 season found that *Star Academy* presented a contemporary version of French culture by drawing upon a common heritage and newer examples of French cultural production. Like *American Idol, Star Academy* presented an idealized version of culture, in which racism does not exist. On *Star Academy*, contestants were constructed as French above all other identity facets—the show constructed "'authentic' representations of French culture" (Darling-Wolf, 2011).

Moreover, the program also portrayed France as a dominant producer in the competitive global-culture marketplace. While *Star Academy* depicted its contestants as knowledgeable French citizens in a global culture, *American Idol* portrayed its contestants as American citizens in a world where seemingly only American culture exists. While France is represented as a "significant global player through the show's meta-narrative" (Darling-Wolf, 2011) other cultures were rarely mentioned on *American Idol*. America was not constructed as a player in the global culture, but rather as the sole producer. In fact, the only other cultures shown in the first seven *American Idol* seasons were poverty-stricken African nations, hardly noted as cultural producers. The charity episodes of *Idol Gives Back* added philanthropy, leadership, sympathy, and generosity characteristics to the American collective identity. Darling-Wolf noted *Star Academy* had its own version of *Idol Gives Back*, in which the show raised money for disaster relief in the People's Republic of Bangladesh. Similar in function to *Idol Gives Back*, this episode built upon French colonialist history and affirmed France's paternalistic position in the world. Through overall show discourse and that particular charity-driven episode, *Star Academy* assured France's benevolence to and importance in the global community. This also occured in America and *Idol Gives Back*. A noticeable difference between *Star Academy* and *American Idol* was *Star Academy's* recognition, at least, that a global culture exists.

In Marwan Kraidy's (2010) recent book, *Reality television and Arab politics: Contention in public life,* he discusses controversies sparked by Arabic reality-TV shows, namely Arabic versions of *Big Brother*, *Star Academy*, and *Superstar* (an officially licensed version of *Pop Idol*). The pan-Arabic programs were comprised of contestants from many countries and generated many questions: Do viewers embrace a multi-national pan-Arabic identity or solely identify with their nation state and its contestants? How can the Western elements present in the shows be reconciled with traditional Islam? Is there a unified Arabic "reality"? What role should women play on reality TV?

While on *American Idol* parity between the sexes was an unstated given, this was not so in Arabic reality TV. Reality TV in Arabic countries pointed out the complex, changing role of women in the contemporary Middle East. As in other genres and places, in Arabic reality TV, portrayal of gender is a "field for the deployment of power" (Kraidy, 2010, p. 197). Female contestants became symbols of nations, market commodities, and representations of changing traditions, as well as jeopardized their safety for seemingly embodying Western values (Kraidy, 2010). Muslim clerics decried the shows' Western values as contrary to their consideration of a virtuous world. Thus, by issuing fatwas and condemnation, "from their perspective, they were resisting the imposition of a foreign reality on their world" (p. 15). While *American Idol* ignored global culture and France's *Star Academy* asserted France's position within it, Arabic reality TV also included elements of global culture, complicating Arabic culture on-screen and offscreen, helping push Arabic countries to modernity. The on-screen reality depicted challenges of traditional culture, often engendering political firestorms (Kraidy, 2010).

Kraidy asserts that different actors "vie to shape social reality"—these actors include government officials, religious leaders, business executives, and activists, among others. Arabic reality TV advances a modern version of reality for Middle East citizens. Reality TV gives Arabs the ability to define their own version of reality, altering the Western-given definition of Arabs (Kraidy, 2010). As Kraidy concludes: "by subjecting the notion of 'reality' to scrutiny, reality TV makes political reality more visible" (p. 198). This, in turn, helps to instigate public debate and in a broadly defined way, democratization. Kraidy finds that accepted democratic elements include social justice and political accountability, while liberal values are rejected. At the same time as this seeming democratization, efforts to control media are also resulting in neo-authoritarianism. Kraidy also notes that Arabic reality-TV competitions are "reaffirming individual nation-states at the expense of pan-Arab identity" (p. 17). For example, Kraidy explains Iraqi citizens rallied around *Star Academy* Iraqi competitor Shadha Hassoun, a woman, which both reaffirmed national identity in the war-torn country and complicated the role of women. However, Kraidy points out that the Arabic national identities are constantly shifting, and older ideas of nationality are consistently remade through reality TV and either the adaptation or dismissal of certain Western ideals.

Just as the French *Star Academy* affirmed French culture and Arabic reality TV facilitated national identities, *American Idol* idealized American identity. *American Idol* represented an image of American culture that is ideally positive and free from controversial race-, gender-, and sexual-orientation equality debates, economic-disparity concerns, and class issues. In *American Idol*, all contestants were positioned as equal despite race, gender, or sexual orientation. Because of this, seemingly everyone can attain the rags-to-riches dream. Equality was an essential characteristic of the *American Idol* ideal. This paralleled the democratic American system of capitalism, where everyone supposedly has the same opportunity to succeed, requiring only hard work. In addition, the show championed the values of hard work, family, talent, and altruism. The *American Idol* American collective identity was fair, optimistic, familial, diligent, and generous.

The message boards also reflected American ideals. As noted previously, communities based around television shows helped to construct the "material and social" conventions of American society (Andrejevic, 2008, 43). On the *American Idol* message boards, fan discourse formed two collective identities: one associated with particular contestants, and the other an overall collective identity associated with being a fan of the show. The distinctive collective identities constructed around individual contestants formed through discourse that included self-identification, inside jokes, relationships, attribution of qualities to the community, and favorable comparison with other communities. The overall *American Idol* fan collective identity formed through discourse emphasizing fan unity and respect, relationships built online that transcended seasons, and shared empathy between contestant groups. Collective identities engendered by community membership created a stronger link for participants to the show itself. Thus, the collective identities facilitated identification with the group and with the show, encouraging potential consumption of *American Idol* and affiliated products. Participants most likely gained pleasure from involvement. Through this involvement a piece of participants' identities were also bound to the show.

All of the collective identities associated with *American Idol*, online and formed through the program discourse, were extremely positive. Typically, on a reality-TV competition, the contestants overtly vie against each other to win. A common refrain heard from contestants on shows other than *American Idol* was "I didn't come here to make friends." On *American Idol*, they seemingly did. When someone was voted off, the other contestants were more likely to cry than exult. Thus, like Cavender's analysis of *Survivor* and *America's Most Wanted*, the discourse analysis of the show found that a traditional conception of community was endorsed through *American Idol*. However, while Cavender (2004) found that discourse-formed reality-TV communities promote a new, negative version of community, I disagree. Quite to the contrary, the *American Idol* communities featured an ideally positive collective identity, free from hatred and aggression. In the *American Idol* communities, as depicted on the show, people were expressly nice to each other, proffering support and empathy. Similarly, the discourse analysis of the *American Idol* message board agrees with this finding. Like Foster's (2004) study of the *Survivor* message boards, the *American Idol* message boards reflected the character of the show. Even the online collective identities, where the us-versus-them comparison portion of social identity theory seemingly expected groups to malign each other, were conspicuously compassionate and polite.

The discourse analysis of the 2008 *American Idol* message boards not only described the existent collective identities but also noted that they mimicked the familial, amiable, and overall nice tone of the show. While groups compared themselves to each other, all fans belonged to the same all-encompassing community in addition to their smaller contestant-based groups. Identification with a community associated with *American Idol* ensured audience loyalty. As a portion of each member's identity was bound to the show, he or she was likely to remain a consumer of the show and related products. Therefore, the message-board-constructed fan collective identities helped assure the continued popularity of *American Idol*, as well as the contemporary social and material status quo.

PRESS COVERAGE: 2002 AND 2008

Additionally, this book also examined contextual *American Idol* press coverage to analyze contemporary cultural opinions of the show. The analysis of these years' press coverage revealed a focus on the economics, finance, and power of the show. Moreover, the two overtones of surprise and derision present in 2002 were absent in 2008. There were two primary conclusions about this finding, both related to the show's popularity. The first hypothesized that journalists were less likely to criticize the show as it grew in popularity. Criticism of the show might have insulted the audience watching it as well. Additionally, there may have been less 2008 criticism due to *American Idol's* lack of novelty. It was old news by 2008. This also explains the lack of surprise related to the program. By 2008, the success of *American Idol* was no longer shocking.

In fact, by 2008 *American Idol* was fully part of American culture. The visibility of *American Idol* in mainstream American culture also explains the lack of contempt in the 2008 media coverage. The popularity of the show guaranteed that show-related news appeared in almost every major news outlet. Stories pertaining to the show permeated contemporary media and therefore contemporary life. Consequently, the ideals within the program had become part of the current American culture as well, from the rags-to-riches theme to the endorsement of beauty ideals. *American Idol* reflected and sustained the status quo of ideals. Journalists may have been hesitant to critique American notions so solidly ingrained in society.

KEY FINDINGS

As demonstrated here, the ideals of celebrity, certain archetypes, and the rags-to-riches theme, among others, were present within *American Idol*. Meehan noted that American "dominant ideology celebrates notions of individuality, freedom, and self-determination. These values are connected to the consumption of goods and services" (Meehan, 2000, p. 75). Values present in *American Idol* may also be connected to maintaining profits for the show. In particular, the show's endorsement of the rags-to-riches narrative signifies a belief in the American system of economics, and culture. *American Idol* has been very financially successful—it behooves the show not to upset the status quo, so as not to diminish profits. The current economic and cultural status facilitated the success of the program. The ideals that *American Idol* illustrated protect the standing of the show while increasing profits. The motives of the show have been benign, in so far as making money is neutral.

While market forces may indeed influence the motives of *American Idol*, the illustration of American ideals cannot be denied. *American Idol's* representation of celebrity supported celebrity as an ideal, reinforced cultural beauty norms, and explained what it means to be a celebrity. Celebrity was constructed as achievable, through lauding average attributes and emotions. Simultaneously, celebrity was constructed as unattainable, as the contestants' celebrity was warranted by innate talent and confirmed through celebrity discourse. Attainable or not, celebrity was a vaunted goal, vehemently desired by the majority of contestants. The achievement of this celebrity by seemingly average people, as demonstrated in discourse asserting contestants are celebrities, made this goal seem easy to achieve, even though talent and hard work were required. Moreover, the discourse of contestants as celebrities was based upon standardized ideals. Discourse stressed contestants should strive to appear glamorously dressed, made-up, and to fit standard weight and body ideals. The judges' comments about who looked like an American idol or a star helped form these celebrity values, particularly Simon's opinions, who often spoke as the voice of the music industry, declaring which contestants would succeed after *Idol.* While the show stressed the current beauty standard for contestants, the audience didn't necessarily heed the show's advice. They voted for whom they liked, thus crowning Ruben Studdard and Taylor Hicks winners, despite the fact that neither looked like a typical celebrity. While winners of the show may not have necessarily fit the established beauty models, such as Ruben Studdard and Taylor Hicks, the judges and overall show discourse stressed the idealized version of celebrity. However, that those winners have not achieved much success outside of the show supported that these show-promoted standards still stand in American culture. Additionally, for every Taylor Hicks there was a winner like Carrie Underwood, beautiful and nice in a conventional way.

Moreover, authenticity is necessary in reality TV to achieve success and celebrity. While the celebrity discourse epitomized the way celebrities appeared and behaved, the archetypal discourse helped fabricate the authenticity of the stars. The discourse-formed archetypes indicated which archetypes people of varied races may inhabit, who may be sexual, and what kinds of sexuality are permitted. Archetypes informed contestants how to act, what to sing, and how to look. In turn, conforming to archetypes was a requirement for authenticity. The authentic portrayal of an archetype helped to foment celebrity. Each archetype, while confining each contestant to certain traits, supposedly matched who the contestants were. Through the archetypes, each contestant was boiled down to an easily digestible persona, familiar and predictable.

Despite the presence of archetypes to which contestants must conform, the show stressed the American value of individuality. The show discourse repeatedly emphasized the ideal of "being yourself." "Being yourself" and being unique, another quality stressed in the show discourse, can also be construed as being an individual. A premise of American capitalism is individual success. Thus, each contestant struggled to differentiate himself or herself. One means of constructing an individual persona is through emotions, which also demonstrate authenticity. Concurrently, emotions made the contestants seem normal and easy to relate to. Whether the contestants truly felt the displayed emotions is debatable. However, the emotions were presented as real, which consequently made the contestants seem sincere. Emotional display is part of the American ideal of individualism and its own ideal through the show's exhibition and praise.

In the past, emotions and the performer's true self were often restricted to "backstage" (Goffman, 1959). On *American Idol*, the backstage, and subsequently emotions, was part of the front stage. Moving the backstage to the front of the stage, while not without historic precedent (for example, the 1950s program *Strike It Rich*), changed *American Idol* from a talent show like *Starsearch* into a combination talent competition/drama. Audiences were able to see the "real," individual selves of the contestants, and why they deserved to be celebrities. Thus, the seemingly authentic and distinctive selves were lauded in addition to contestants' talent.

On *American Idol*, celebrity achievement was constructed to be as American as the existent democratic system. On the show, celebrity and success was achieved through the democratic voting process, a parallel of the American system of political election. Just as the American democratic system outwardly gives individual voters agency, so did *American Idol*. Through the mimicry of the democratic voting system, *American Idol* endorsed the American election process. While *American Idol* never revealed the exact number of how many votes each contestant received, it still supported the voting system. *American Idol* naturalized and helped sustain belief in the value of the existent voting system, allowing no other election means to exist. Perhaps *American Idol's* direct voting system may even be adapted in the future for political elections. The *American Idol* audience will already be accustomed to it.

As Lule (2001) asserted, myths are ways society transmits ideology and values. *American Idol* is a contemporary myth, rife with meaning. On the show, the American collective identity was depicted as familiar and congenial (particularly in contrast to Simon Cowell). The program's archetypes were fun, caring, and hopeful despite their hard-luck beginnings. The discourse of dreams gave audience members something to dream of attaining. Watching an easy-to-relate-to and identifiable person achieve success through the show might have caused a heartwarming and joyful vicarious emotion—or jealousy. Just as traditional myths may encourage the extant social order (Lule, 2001) so did *American Idol*. *American Idol* naturalized voting, individualism, and the ideals of hard work and equal opportunity for fame. In general, the rags-to-riches narrative encourages hard work in order to receive rewards. The reality-television version promises quick fame. However, *American Idol* equally stressed working and talent. Effort, skill, and authenticity were required to achieve success on *American Idol* and subsequent celebrity. In addition to those requirements, a suitable background, easily identifiable archetype, and affable nature were necessary to be an *Idol* celebrity.

The America constructed on *American Idol* is a complication-free, perfect place where democracy works flawlessly and everyone has an equal opportunity to achieve his or her American rags-to-riches dream. Through the show discourse, *American Idol* also established its audience as a single, collective entity. This group of people shared two collective identities: the first as fans of the show and the second as Americans. The show discourse repeatedly labeled the audience as American and employed various rhetorical techniques to reinforce this identity. The *American Idol* discourse constructed an America composed of the *American Idol* audience, fans that embodied the show's discursive ideals. The analysis of the fan message-board discourse supported the idea that fans often absorb and reiterate the show's ideas, such as the importance of congeniality. The show itself often literally called the audience *America*. As Ryan Seacrest reminded those at home, "America, you decide." The audience was assumed to have absorbed the messages of the show, the ideals of celebrity, authenticity, individuality, family, beauty, sexuality, consumption, democracy, and to have applied them in choosing a winner, their idol. The winner, in turn, also represented the *American Idol* ideals.

CONTRADICTIONS

Throughout the evidence in this book, several contradictions have appeared. They include dichotomies of nice/nasty, humility/confidence, archetype/individualism, community/individualism, and celebrity as intimate/celebrity on a pedestal. These deserve further exploration as they complicate several of the noted ideals.

First, there is a blatant contradiction between the overall nice, friendly, and amiable tone of the show and Mr. Mean, Simon Cowell. The good-natured atmosphere of the show, visible in the contestants' camaraderie and discursive emphasis on compassion and fun, is a sharp contrast to Simon's cutting criticism. While the contestants are repeatedly confirmed to be celebrities through a variety of means, Simon and occasionally other judges remark on the contestants' lack of skill during performance critiques. While Simon's commentary is important to the show (indeed, it may prove to be a main element attracting an audience) the heavy amount of celebrity discourse outweighs his comments. In other words, the celebrity confirmation on the show is almost oppressive in its presence; Simon's negativity cannot lessen it or its effects. Unless, of course, Simon's comments caused a contestant to be eliminated, which cannot be proven. This contradiction between the family-friendly, nice tone of the show and the nastiness of Simon Cowell serves to point out the intensity of each other. The show is overwhelmingly nice; Cowell is overwhelmingly nasty. This dichotomy functions to highlight each and possibly helps to fascinate the audience.

Another interesting contraction is between the conflicting personality traits of humility and confidence. While these seem to be oppositional qualities, in fact, *American Idol* asserts they must be present within the same person and often exhibited simultaneously, which would seem to be impossible if it wasn't accomplished so frequently on the show. Contestants are exhorted again and again to be confident, to express without fear whom each truly is. A concurrent theme is humility. While no one on the show expressly uttered the words, "do not be arrogant," the importance of being liked implied this. Being likable was very much emphasized as a necessary quality (presumably in order to receive votes and in turn establish celebrity). Humility could be part of trying to be liked—its opposite, self-importance, is certainly a part of being disliked. A notable example of a person who exhibited both confidence and humility on the show, often in the same performance, was season-seven runner-up David Archuleta. David's performances were usually flawless and well-received by the judges and fans, indicated by cheering and applause. He performed with confidence as well; he rarely made a noticeable mistake, never had a nervous quaver in his voice, and always used firm and decisive body language. He never conveyed any nervousness or anxiety immediately before, during, or after a performance. Yet, he always seemed humbly surprised upon receiving cheers and praise. He awkwardly accepted compliments, smiled graciously, and outwardly often lacked words to express his gratitude, although he usually managed to eke out a "thank you." David was both humble and confident.

The humility/confidence paradox is related to the dichotomy of unreachable celebrity and intimate celebrity. Often, a person deserves celebrity because of innate talent or merit, which in turn begets confidence. Confidence engenders a celebrity that is not a regular person, but rather is somehow better than the audience. On the other hand, humility and modesty create an intimate celebrity, a TV celebrity audiences are intimately familiar with (Marshall, 1997) and one who may be interchangeable with the audience member (Holmes, 2004a). Humility and confidence, along with many other factors, create different types of celebrity. The *American Idol* celebrity is thus a music celebrity, distant and with a definable talent, separated from the masses through technology (Marshall, 1997) and a TV celebrity, intimately known (Holmes, 2004; Marshall, 1997). The *American Idol* celebrity is both personally known to viewers though backstage access and emotional displays, yet also is a distant music star, with a definable talent that most people cannot match. The *American Idol* celebrity is a distinctive combination of unattainable stardom and every-day familiarity. This parallels the combination of celebrity confirmation and authenticity. Innate talent merits celebrity status, creating a celebrity on a pedestal, better than the average person. Authenticity helps create the intimate celebrity, someone with whom the audience can relate.

Another facet of *American Idol* celebrity construction is the contradiction between archetypes and individualism, both themes stressed within the show. Contestants must conform to an archetype to achieve celebrity, yet also express their individuality. Contestants are (seemingly willingly) placed into archetypes through discourse, including being labeled by the judges and host, musical choices, and sartorial style. Contestants who do not adhere to an archetype are condemned as not knowing who they are or being confusing. While conformity is a requirement for the contestants, individuality is stressed. Contestants are continually told to "be themselves" and the discourse stresses that each could triumph due to individual skill. The contestants walk a fine line of emulating those who came before them in their archetype and trying to express individuality. This is accomplished by fitting a general archetype, yet modifying it a little. For example, the program's discourse affirmed Chris Daughtry fit the rock archetype. Yet he also managed to demonstrate his personal take on the archetype by adding the component of fatherhood. His stepchildren were often featured on the show and Chris frequently spoke of his love for his family. Thus, typically successful contestants fit an archetype, though slightly altered it, simultaneously conforming and expressing individuality. This emphasis on individuality and individual success (pulling herself up by her bootstraps, achieving alone) ignores the community formed around the contestant, voting and rooting for her. On the show, individual work and performance is stressed much more than a collective effort to attain success. This presents a secondary contradiction between community and individualism.

On *American Idol* both community and individualism are lauded, complicating each. Just as contestants are asked to be themselves, the audience is asked to participate by voting. As chapter seven noted, the discourse of the show forms a community of the audience. Not only is the large audience constructed as a collective American group, but the show discourse forms small groups as well, regional and contestant-based. Similar collective groups form online, as chapter eight discussed. How are these communities reconciled with the show's individual emphasis? Seemingly oppositional, they coexist, helping promote celebrity. The contestants seem to expect to be regarded as exceptional and accepted as celebrities through the emphases on individualism and celebrity. The audience is also groomed to be receptive to this celebrity through celebrity and individualism discourse. Moreover, the community formation creates a built-in audience. Individual celebrities do not exist without an audience, a watching and consuming community. This duality of community/individualism also promotes the current cultural and economic status quos. *American Idol* individualism buttresses the American dream, while community provides the means for popularity.

These inherent contractions present on the show complicate many of the ideals explicated in this book. However, this does not negate their presence or importance—the contradictions on the show reflect the dissonance people face on a daily basis. These contradictions also represent the mixture of diverse people in America and on the program, and this mix of cultures results in sometimes contradictory American ideals.

SUMMARY

This research stems from a particular class, gender, and race perspective. Despite academic training, as a white female I may be predisposed to notice particular information and may not detect everything pertinent. Nevertheless, I contend this interpretation of the research is valid, due to the amount of meticulous analysis and interpretation. This book contextualizes *American Idol's* place in academic research by combining show discourse, an online fan reception study, and a press-coverage study. This methodology is a triangulated approach, taking different forms of media into account, yet deriving similar results from each analysis. This makes the findings significant. I found themes of a rags-to-riches narrative, archetypes, and celebrity importance, as well as a focus on congenial collective identities, in the show's press coverage, online message boards, and throughout seven seasons of the show itself. Thus, this research occupies a unique place in academia, linking *American Idol*, television, celebrity, collective identity, and new media studies. It fills a gap in the existent literature concerning reality television, new media, and celebrity.

This work adds significantly to varied bodies of literature. It contributes a media perspective to the American studies literature. To the celebrity literature, this book adds an in-depth study of the conflicted nature of the authentic celebrity creation process. Additionally, the press and online reception studies include audience members' understanding of the show and its celebrity construction. For the collective identity literature, this project analyzes how the show foments various collective identities as well as how the online fan communities form. Finally, this research adds a comprehensive analysis of the most popular television program of the first decade of the 2000s to the television and culture literature.

Through the combination of program discourse analysis, message board study, and press coverage analysis, this study was able to make claims about the celebrity, collective identity, and American ideals that no other study has previously discerned. This study is limited in that it only includes one program, examined just the first seven seasons of *American Idol*, the message board study included one season, and the press coverage studies covered only two years. Future research on this topic could expand the study to include more seasons of *American Idol*, more seasons of the *American Idol* message boards, message boards from other reality television shows, other online fan material such as blogs and unofficial fan message boards, or other national versions of *Idol*. Comparisons between *American Idol* and other countries' similar programming may reveal cultural differences and similarities, as well as demonstrate global economic and cultural flows. Future research could examine if the format replication in other markets contains corresponding national elements and/or what indigenous values are present.

Through this work, it becomes clear that *American Idol* presents an idealistic vision of American life, where everyone is equal, democracy chooses the right winner, power is shared, and celebrity and success can be attained simply by "being yourself." In *American Idol's* rendering of America, racism, homophobia, xenophobia, and sexism do not exist. This idealized America consists of the ideological systems and apparatus already in place. Through *American Idol's* American themes, representations of celebrity, and facilitation of collective identities, the show provides an idealized version of American culture. These idealized versions of America, national identity, and celebrity support contemporary economic and cultural norms: capitalistic ideology and concomitant materialism, beauty, race, gender, and sexuality standards, and ostensible equality epitomized by possibility for success.

References

Abrams, D. (1996). Social identity, self as structure and self as process. In P. Robinson (Ed.), *Social groups and identities: Developing the legacy of Henri Tajfel* (143–168). Boston, MA: Butterworth-Heinemann.

Adalian, J. (2002, July 23). NBC rewrites summer. *Daily Variety*. Retrieved August 1, 2009, from the Lexis Nexis database.

Alberoni, F. (1962/1972). The powerless elite. In D. McQuail (ed.), *Sociology of mass communications* (75–98). Middlesex, UK: Penguin Books, Ltd.

Althusser, L. (1971). *Lenin and philosophy and other essays*. New York, NY: Monthly Review Press.

American Idol is going to Disney World! Press release. Retrieved February 7, 2008, from biz.yahoo.com/prnews/080207/clth126.html?v=41.

American Idol puts its tunes on iTunes. (2008, February 20). *The St. Petersburg Times (Florida)*. Retrieved August 1, 2009, from the Lexis Nexis database.

Anderson, B. (1983/2003). *Imagined communities: Reflections on the origin and spread of nationalism*. New York, NY: Verso.

Andrejevic, M. (2002). The kinder, gentler gaze of Big Brother: Reality TV in the era of digital capitalism. *New Media and Society* 4(2), 251–270.

Andrejevic, M. (2004). *Reality TV: The work of being watched*. New York, NY: Rowman & Littlefield Publishers, Inc.

Andrejevic, M. (2008). Watching television without pity: The productivity of online fans. *Television & New Media* 9 (1), 24–46.

Ang, I. (1985). *Watching 'Dallas': Soap opera and the melodramatic imagination*. London, UK: Methuen.

Atkinson, C. (2007, January 8). 'Idol' juggernaut passes $2.5 bil and hits the gas. Advertising Age, 78 (2), 1–29. Retrieved April 16, 2007, from the Academic Search Premier database.

Bagdikian, B. (2004). *The new media monopoly*. Boston, MA: Beacon Press.

Barker, C. (1997). *Global television: An introduction*. Oxford, UK: Blackwell.

Barnfield, G. (2002). From direct cinema to car-wreck video: Reality TV and the crises of content. In D. Cummings (Ed.), *Reality TV: How real is real?* (47–66). Oxford, UK: Hodder & Stroughton.

Barthes, R. (1957/1995). *Mythologies* (Trans. A. Lavers). New York. NY: Hill and Wang.

Barthes, R. (1977). Introduction to the structural image of the narrative. In *Image, Music, Text* (79–124). New York, NY: Hill and Wang.

Battaglio, S. (2002a, June 10). Say hi to "Mr. Nasty": Idol features vicious judge. *Daily News (New York),* 84. Retrieved August 1, 2009, from the Lexis Nexis database.

Battaglio, S. (2002b, June 24). New brand of reality. *Daily News* (New York), 78. Retrieved August 1, 2009, from the Lexis Nexis database.

Baym, N. (1998). Talking about soaps: Communicative practices in a computer mediated fan culture. In C. Harris and A. Alexander (Eds.). *Theorizing fandom: Fans, subculture and identity* (111–310). Cresskill, NJ: Hampton Press Inc.

Bazalgette, P. (2001, October). *Big Brother* and beyond—Golden age be damned. *Television,* 20–23.

Benjamin, W. (1936/1973).The work of art in the age of mechanical reproduction. In H. Arendt (Ed.), *Illuminations* (211–45). London, UK: Fontana.

Bianco, R. (2002, August 19). New reality is a lot nastier. *USA Today.* Retrieved August 1, 2009, from the Lexis Nexis database.

Bignell, J. (2005). *Big Brother: Reality TV in the twenty-first century.* New York, NY: Palgrave Macmillan.

Billig, M. (1995). *Banal nationalism.* Thousand Oaks, CA: Sage.

Bird, E., and Dardenne, R. (1988). Myth, chronicle, and story: Exploring the narrative qualities of news. In J. Carey (Ed.), *Media, myths, and narrative* (67–86). Beverly Hills, CA: Sage Publications.

Bird, E., and Dardenne, R. (1997). Myth, chronicle, and story: Exploring the narrative qualities of news. In D. Berkowitz (Ed.), *Social meanings of news* (333–350). Thousand Oaks, CA: Sage.

Biressi, A., and Nunn, H. (2005). *Reality TV: Realism and revelation.* New York, NY: Wallflower Press.

Boorstin, D. (1961/1992). *The image.* New York, NY: Vintage Books.

Braudy, L. (1997). *The frenzy of renown.* New York, NY: Vintage Books.

Britt, D. (2002, September 6). Fox's 'Idol': American in so many ways. *The Washington Post.* Retrieved August 1, 2009, from the Lexis Nexis database.

Brooker, W. (2002). Overflow and audience. In W. Brooker and D. Jermyn (Eds.), *The audience studies reader* (322–33). London, UK: Routledge.

Brooks, T., and Marsh, E. (2003). *The complete directory to prime time network and cable TV shows, 1946-present.* New York, NY: Ballantine Books.

Brune, R., and Ross, S. (Executive Producers). (2002a, June 18). *Good Morning America* [Television Broadcast]. New York, NY: ABC. Transcript retrieved August 1, 2009, from the Lexis Nexis database.

Brune, R., and Ross, S. (Executive Producers). (2002b, September 5). *Good Morning America* [Television Broadcast]. New York, NY: ABC. Transcript retrieved August 1, 2009, from the Lexis Nexis database.

Brune, R., Ross, S., and Sherwood, B. (Executive Producers). (2008, March 14). *Good Morning America* [Television Broadcast]. New York, NY: ABC. Transcript retrieved August 1, 2009, from the Lexis Nexis database.

Burke, P. (1992). *The fabrication of Louis XIV*. New Haven, CT: Yale University Press.

Caldwell, W. (2006). *Cynicism and the evolution of the American dream*. Washington, DC: Potomac Books, Inc.

Carter, B. (2002, July 16). Dreaming up summer's breed of TV hits. *The New York Times*. Retrieved August 1, 2009, from the Lexis Nexis database.

Cassidy, M. (2001). Visible storytellers: Women narrators on 1950s daytime television. *Style, 35* (2), 354–374.

Cathcart, R. (1994). From hero to celebrity: The media connection. In S. J. Drucker and R. S. Cathcart (Eds.), *American heroes in a media age* (36–46). Cresskill, NJ: Hampton Press, Inc.

Cavender, G. (2004). In search of community on reality TV: *America's Most Wanted* and *Survivor*. In S. Holmes, and D. Jermyn (Eds.), *Understanding reality television* (154–172). New York, NY: Routledge.

Chatman, S. (1981). What novels can do that films can't (and vice versa). In W. J. T. Mitchell (Ed.), *On narrative* (117–36). Chicago, IL: University of Chicago Press.

Chiricos, T., and Eschholz, S. (2002). The racial and ethnic typification of crime and the criminal typification of race and ethnicity in local television news. *Journal of Research in Crime and Delinquency, 39* (4), 400–20.

Clarke, M. (1987). *Teaching popular television*. London, UK: Methuen.

Clerc, S. (1996). DDEB, GATB, MPPB, and ratboy: The X-Files media fandom, online and off. In D. Lavery, A. Hague, and M. Cartwright (Eds.), *Deny all knowledge: Reading the X-Files* (36–51). Syracuse, NY: Syracuse University Press.

Clissold, D. (2004). Candid Camera and the origins of reality TV: Contextualizing a historical precedent. In S. Holmes and D. Jermyn (Eds.), *Understanding reality television* (33–53). New York: Routledge.

Coles, G. (2000). "Docusoap": Actuality and the serial format. In B. Carson and M. Llewllyn-Jones (Eds.), *Frames and fictions on television: The politics of identity in drama* (27–39). Wiltshire, UK: Cromwell Press.

Cormack, M. (1995). *Ideology*. Ann Arbor, MI: University of Michigan Press.

Couldry, N. (2003). *Media rituals: A critical approach*. New York, NY: Routledge.

Couldry, N. (2004). Teaching us to fake it: The ritualized norms of television's "reality" games. In S. Murray and L. Ouellette (Eds.), *Reality TV: Remaking television culture* (57–74). New York, NY: New York University Press.

Creamer, J. (2001, September). Chain of events. *Televisual*, 11–13.

Crosbie, L. (2008, March 4). No one can be authentic on a show this rigid. *The Globe and Mail (Canada)*. Retrieved August 1, 2009, from the Lexis Nexis database.

Cullen, J. (2003). *The American dream: A short history of an idea that shaped a nation*. New York, NY: Oxford University Press.

Dahlgren, P. (2000). Key trends in European television. In J. Wieten, P. Dahlgren, and G. Murdock (Eds.), *Television across Europe: A comparative introduction* (23–35). London, UK: Sage.

Darling-Wolf, F. (2011). World citizens 'a la Française': *Star Academy* and the negotiations of 'French' identities. In M. Kraidy and K. Sender (Eds.), *The politics of reality television: Global perspectives* (127–39). New York, NY: Routledge

Dauncey, H. (1996). French reality television. *European Journal of Communication*, 11, 83–106.

de Cordova, R. (1990). *Picture personalities: The emergence of the star system in America*. Champaign, IL: University of Illinois Press.

de la Fuente, A. (2008, March 31). FremantleMedia fires up 'Idol' in Brazil. *Daily Variety*. Retrieved August 1, 2009, from the Lexis Nexis database.

de Moraes, L. (2002, September 4). September 11: *American Idol* Seizes the Day. *The Washington Post*. Retrieved August 1, 2009, from the Lexis Nexis database.

de Moraes, L. (2008, May 22). 'American Idol' has a winner: It's . . . David. *The Washington Post*. Retrieved August 1, 2009, from the Lexis Nexis database.

Deggans, E. (2002, September 3). 'Idol' curiosities. *The St. Petersburg Times (Florida)*. Retrieved August 1, 2009, from the Lexis Nexis database.

della Cava, M. (2008, April 17). End of 'Rainbow' is nowhere in sight: Hawaiian version gets 'Idol' boost. *USA Today*. Retrieved August 1, 2009, from the Lexis Nexis database.

DeLuca, D. (2002, June 18). Brit shows no mercy to singing hopefuls. *The Philadelphia Inquirer*. Retrieved August 1, 2009, from the Lexis Nexis database.

Derrida, J. (1981). The law of genre. In W. J. T. Mitchell. (Ed.), *On Narrative* (51–78). Chicago, IL: University of Chicago Press.

Douthit, R., and Whitworth, W.W. (Executive Producers). (2008, April 7). *Larry King Live* [Television Broadcast]. Atlanta, GA: CNN. Transcript retrieved August 1, 2009, from the Lexis Nexis database.

Dovey, J. (2000). *Freakshow: First person media and factual television*. London: Pluto.

Dow, B. (1996). Prime-time feminism: Television, media culture, and the women's movement since 1970. Philadelphia, PA: University of Pennsylvania Press.

Dyer, R. (1986/2004). *Heavenly bodies: Film stars and society*. (2nd edition). New York, NY: Routledge.

Dyer, R. (1998). *Stars*. London, UK: BFI Publications.

Dyer, R. (2000). The role of stereotypes. In P. Marris and S. Thornham (Eds.), *Media studies: A reader* (245–51). New York, NY: New York University Press.

Ellis, J. (1992). *Visible fictions: Cinema, television, video*. London, UK: Routledge.

Etzioni, A. (1993). *The spirit of the community: Rights, responsibilities, and the communitarian agenda*. New York, NY: Crown.

Evans, J. (2005a). Celebrity: What's the media got to do with it? In J. Evans and D. Hesmondhalgh (Eds.), *Understanding media: Inside celebrity* (1–10). Berkshire, UK: Open University Press.

Evans, J. (2005b). Celebrity, media and history. In J. Evans and D. Hesmondhalgh (Eds.), *Understanding media: Inside celebrity* (11–56). Berkshire, UK: Open University Press.

Fairchild, C. (2007). Building the authentic celebrity: The 'Idol' phenomenon in the attention economy. *Popular Music & Society*, 30 (3), 355–75.

Fairclough, N. (2000). Critical analysis of a media discourse. In P. Marris and S. Thornham (Eds.), *Media studies: A reader* (308–25). New York, NY: New York University Press.

Fairclough, N. (2003). *Analyzing discourse: Textual analysis for social research.* New York: Routledge.

Feuer, J. (1986). Narrative form in American network television. In C. MacCabe (Ed.), *High theory/Low culture* (101–14). Manchester, UK: Manchester University Press.

Fink, S. (2008, February 16). Get over the pros. *The St. Petersburg Times (Florida).* Retrieved August 1, 2009, from the Lexis Nexis database.

Fishman, M., and Cavender, G. (Eds). (1998). *Entertaining crime: Television reality programs.* New York, NY: Aldine de Gruyter.

Fiske, J. (1987). *Television culture.* New York, NY: Methuen.

Fiske, J. (2000). The codes of television. In P. Marris and S. Thornham (Eds.), *Media studies: A reader* (220–30). New York, NY: New York University Press.

Foster, D. (2004). "Jump in the pool": The competitive culture of *Survivor* fan networks. In S. Holmes and D. Jermyn (Eds.), *Understanding reality television* (270–89). New York, NY: Routledge.

FOX Interactive Media. Retrieved April 18, 2007 from songwriter.americanidol.com/theSearch/.

FremantleMedia. Retrieved April 16, 2007 from www.fremantlemedia.com.

Frere-Jones, S. (2008, May 19). Idolatry: Everybody's a critic. *The New Yorker.* Retrieved August 1, 2009, from the Lexis Nexis database.

Friedman, J. (2002). Introduction. In J. Friedman (Ed.), *Reality squared* (1–22). New Brunswick, NJ: Rutgers University Press.

Friedman, S. (Executive Producer). (2002a, September 5). *The Early Show* [Television Broadcast]. New York, NY: CBS. Transcript retrieved August 1, 2009, from the Lexis Nexis database.

Friedman, S. (Executive Producer). (2002b, September 9). *The Early Show* [Television Broadcast]. New York, NY: CBS. Transcript retrieved August 1, 2009, from the Lexis Nexis database.

Fuller bids for CKX takeover (2010, June 1). Music Week. Retrieved April 19, 2011 from www.musicweek.com/story.asp?storycode=1041351.

Fuller to manage American Idol winner (2011, March 29). Music Week. Retrieved April 19, 2011 from www.musicweek.com/story.asp?sectioncode=1&storycode=1044737&c=1.

Gamson, J. (1994). *Claims to Fame: Celebrity in contemporary America.* Berkeley, CA: University of California Press.

Gans, H. (1974/2004). *Deciding what's news.* Evanston, IL: Northwestern University Press.

Gardner, E., and Keveney, B. (2002, July 30). Fox hit 'American Idol' mixes music, meanness. *USA Today.* Retrieved August 1, 2009, from the Lexis Nexis database.

Garofalo, R. (1994). Culture versus commerce: The marketing of Black popular music. *Public Culture,* 7(1), 275–88.

Garrity, B., and Waddell, R. (2007, January 20). The 'American' Way. *Billboard*, 119 (3), 21–23. Retrieved April 14, 2007, from the Academic Search Premier database.

Geraghty, C. (2000). Re-examining stardom: questions of texts, bodies, and performance. In C. Gledhill and L. Williams (Eds.), *Reinventing Film Studies* (183–201). London, UK: Arnold.

Gilbert, M. (2002, September 4). 'Idol' has the crass to make it must-see. *The Boston Globe*. Retrieved August 1, 2009, from the Lexis Nexis database.

Gillin, B. (2002, September 1). Why 'Idol' is getting America's votes. *The Philadelphia Inquirer*. Retrieved August 1, 2009, from the Lexis Nexis database.

Gitlin, T. (1992). Blips, bites, and savvy talk: Television's impact on American politics. In D. Shimkin, H. Stolerman, and H. O'Connor (Eds.), *State of the art: Issues in contemporary mass communication* (213–22). New York, NY: St. Martin's.

Goffman, E. (1959). *The presentation of self in everyday life*. Garden City, NY: Doubleday & Co.

Grady, T. (Producer). (2002, July 24). *CNN Live Today* [Television Broadcast]. Atlanta, GA: CNN. Transcript retrieved August 1, 2009, from the Lexis Nexis database.

Gundersen, E. (2002, July 25). RCA hopes 'Idol' tunes spin gold. *USA Today*, 1D. Retrieved August 1, 2009, from the Lexis Nexis database.

Gundersen, E. (2008, May 21). 'Time' is now for winner of Idol songwriter contest: 'Housedad' crafts inspirational song. *USA Today*. Retrieved August 1, 2009, from the Lexis Nexis database.

Hall, S. (1984). The narrative construction of reality. *Southern Review,* 17 (1), 3–17.

Halperin, S. (2008). Taylor Hicks, Ruben Studdard dropped from J Records? *Entertainment Weekly*. Retrieved February 7, 2008 from hollywoodinsider.ew.com/2008/01/two-idol-winner.html.

Hay, C. (2002a, June 2). Music and showbiz. *Billboard*, 64. Retrieved August 1, 2009, from the Lexis Nexis database.

Hay, C. (2002b, August 3). 'American Idol' weds reality TV and music. *Billboard*, 65. Retrieved August 1, 2009, from the Lexis Nexis database.

Heller, K. (2002, August 18). Screen savers: There are really, really bad TV shows to savor. And the Tony-est one to anticipate. *The Philadelphia Inquirer*. Retrieved August 1, 2009, from the Lexis Nexis database.

Hill, A. (2002). *Big Brother*: The real audience. *Television and New Media* 3 (3), 323–340.

Hill, A. (2005). *Reality TV: Audiences and popular factual television*. New York, NY: Routledge.

Hill, A. (2007). *Big Brother*: The real audience. In H. Newcomb (Ed.), *Television: The critical view (*7th ed*)* (471–485). New York, NY: Oxford University Press.

Hiltbrand, D. (2002, September 1). The 'Idol' advantage. *The Philadelphia Inquirer*. Retrieved August 1, 2009, from the Lexis Nexis database.

Hoffman, S. (Executive Producer). (2008a, January 16). *The Bryant Park Project* [Radio Broadcast]. New York, NY: NPR. Transcript retrieved August 1, 2009, from the Lexis Nexis database.

Hoffman, S. (Executive Producer). (2008b, March 11). *The Bryant Park Project* [Radio Broadcast]. New York, NY: NPR. Transcript retrieved August 1, 2009, from the Lexis Nexis database.

Hoffman, S. (Executive Producer). (2008c, May 9). *The Bryant Park Project* [Radio Broadcast]. New York, NY: NPR. Transcript retrieved August 1, 2009, from the Lexis Nexis database.

Hogg, M. (1996). Intergroup processes, group structure and social identity. In P. Robinson (Ed.), *Social groups and identities: Developing the legacy of Henri Tajfel* (65–94). Boston: Butterworth-Heinemann.

Holmes, S., and Jermyn, D. (2004). Introduction: Understanding reality TV. In S. Holmes and D. Jermyn (Eds.), *Understanding reality television* (1–32). New York, NY: Routledge.

Holmes, S. (2004a). "All you've got to worry about is the task, having a cup of tea, and doing a bit of sunbathing": Approaching celebrity in *Big Brother*. In S. Holmes and D. Jermyn (Eds.), *Understanding reality television* (111–35). New York, NY: Routledge.

Holmes, S. (2004b). "Reality goes pop!": Reality TV, popular music, and narratives of stardom in *Pop Idol. Television and New Media,* 5 (2), 147–72.

Horkheimer, M., and Adorno, T. (1972). *Dialectic of enlightenment.* (Trans. J. Cumming,). New York, NY: Herder and Herder. (Original work published 1944).

Howard, T. (2002, September 9). Real winner of 'American Idol': Coke. *USA Today.* Retrieved August 1, 2009, from the Lexis Nexis database.

James, C. (2002, September 1). On 'Idol,' the only losers are the audience's ears. *The New York Times.* Retrieved August 1, 2009, from the Lexis Nexis database.

Jenkins, H. (1988). *Star Trek* rerun, reread, rewritten: Fan writing as textual poaching. *Critical Studies in Mass Communication,* 5 (2), 85–107.

Jenkins, H. (2001). Convergence? I diverge. *Technology Review,* 104 (5), 93.

Jenkins, H. (2006). *Convergence culture: Where old and new media collide.* New York, NY: New York University Press.

Jermyn, D. (2004). "This is about real people!": Video technologies, actuality and affect in the television crime appeal. In S. Holmes and D. Jermyn (Eds.), *Understanding reality television* (71–90). New York, NY: Routledge.

Kammen, M. (1978). *A season of youth: The American Revolution and the historical imagination.* New York, NY: Knopf.

Kaplan, R. (Executive Producer). (2008, May 21). *CBS Evening News with Katie Couric.* [Television Broadcast]. New York, NY: CBS. Transcript retrieved August 1, 2009, from the Lexis Nexis database.

Katz, E., and Liebes, T. (1990). *The export of meaning: Cross-cultural readings of Dallas.* New York: Oxford University Press.

Kavka, S., and West, A. (2004). Temporalities of the real: Conceptualizing time in reality TV. In S. Holmes and D. Jermyn (Eds.), *Understanding reality television* (136–53). New York, NY: Routledge.

Kavka, M. (2005). Love 'n' the real: Or, how I learned to love reality TV. In G. King (Ed.), *The spectacle of the real: From Hollywood to reality TV and beyond* (93–104). Portland, OR: Intellect Books.

Kermode, F. (1981). Secrets and narrative sequence. In W. J. T. Mitchell (Ed.), *On narrative* (79–98). Chicago, IL: University of Chicago Press.

Keveney, B. (2002a, June 18). Viewers 'Idol'-ize this saucy search for a superstar. *USA Today*. Retrieved August 1, 2009, from the Lexis Nexis database.

Keveney, B. (2002b, September 4). 'Idol' worship: It's not over yet. *USA Today*, 1D. Retrieved August 1, 2009, from the Lexis Nexis database.

Keveney, B. (2006, January 19). 35.5M viewers audition fifth-season 'Idol.' *USA Today*. Retrieved April 14, 2007, from the Academic Search Premier database.

Kilborn, R. (1994). "How real can you get": Recent developments in "reality television." *European Journal of Communications*, 9, 421–39.

Kinon, C. (2008, February 22). They're music to her ears: Abdul has faves she 'Idol'-izes. *Daily News (New York)*. Retrieved August 1, 2009, from the Lexis Nexis database.

Kitch, C. (2005). *Pages from the past: History and memory in American magazines*. Chapel Hill, NC: University of North Carolina Press.

Kompare, D. (2004). Extraordinarily ordinary: *The Osbournes* as "An American family." In S. Murray and L. Ouellette (Eds.), *Reality TV: Remaking television culture* (97–116). New York, NY: New York University Press.

Kraidy, M. (2006). Reality television and politics in the Arab world (preliminary observations). *Transnational Broadcasting Studies*, 2(1), 7–28.

Kraidy, M. (2010). *Reality television and Arab politics: Contention in public life*. New York, NY: Cambridge University Press.

Kuklick, B. (1972/1999). Myth and symbol in American studies. In L. Maddox (Ed.), *Locating American studies: The evolution of a discipline* (71–90). Baltimore, MD: The Johns Hopkins University Press.

Langer, J. (1981). TV's personality system. *Media, Culture and Society*, 4, 351–65.

Leishman, F., and Mason, P. (2002). *Policing and the media: Facts, fictions, and factions*. Cullompton, UK: Willan Publishing.

Levine, D. (Executive Producer). (2008, January 16). *Showbiz Tonight* [Television Broadcast]. Atlanta, GA: CNN. Transcript retrieved August 1, 2009, from the Lexis Nexis database.

Lieberman, D. (2007, April 9). Fans eat up 'Idol' merchandise, and wear it and so . . . *USA Today*. Retrieved April 14, 2007, from the Academic Search Premier database.

Lieberman, D. (2008, May 19). Fat lady hasn't sung for 'Idol': Fox's talent show still going strong despite ratings slip. *USA Today*. Retrieved August 1, 2009, from the Lexis Nexis database.

Lippmann, W. (1922). *Public Opinion*. New York: Harcourt Brace.

Lipsitz, G. (1990). *Time passages: Collective memory and American popular culture*. Minneapolis, MN: University of Minnesota Press.

Lisotta, C. (2006, February 13). *Why 'Idol' is golden. Television Week*, 25 (7). Retrieved April 14, 2007, from the Academic Search Premier database.

Lotz, A. (2006). *Redesigning women: television after the network era.* Urbana, IL: University of Illinois Press.

Lule, J. (2001). *Daily news, eternal stories: The mythological role of journalism.* New York, NY: Guilford Press.

Lythgoe, N. (Producer). (2002–2008). *American Idol* [Television series] Los Angeles, CA: Fox.

Madger, T. (2004). The end of TV 101: Reality programs, formats and the new business of television. In S. Murray and L. Ouellette (Eds.), *Reality TV: Remaking television culture* (137–156). New York, NY: New York University Press.

Magubane, Z. (2001). Which bodies matter? Feminism, poststructuralism, race, and the curious theoretical odyssey of the "Hottentot Venus." *Gender and Society,* 15 (6), 816–34.

Marking, H. (Director). (2009). *Afghan Star.* Kabul, AF: Kaboora Production.

Marris, P. and Thornham, S. (2000). Textual structure. In P. Marris and S. Thornham (Eds.), *Media studies: A reader* (195–99). New York, NY: New York University Press.

Marshall, P. D. (1997). *Celebrity and power: Fame in contemporary culture.* Minneapolis, MN: University of Minnesota Press.

Mason, M. S. (2002, August 9). 'Idol' is burnished, but still tarnished. *Christian Science Monitor.* Retrieved August 1, 2009, from the Lexis Nexis database.

Maul, K. (2008, April 21). McDonald's teams with 'Idol.' *PR Week.* Retrieved August 1, 2009, from the Lexis Nexis database.

Mayfield, G. (2008, March 22).Over the counter: 'Idol' fans sing a loud 'Hallelujah.' *Billboard,* 65. Retrieved August 1, 2009, from the Lexis Nexis database.

Meehan, E. (2000). Introduction: Consuming audiences? Production and reception in media research. In I. Hagen and J. Wasko (Eds.), *Consuming audiences? Production and reception in media research* (71–92). Cresskill, NJ: Hampton Press Inc.

McGinn, A. (2008a, April 16). She stuck to her faith and to her music; Carmen Rasmusen. *USA Today.* Retrieved August 1, 2009, from the Lexis Nexis database.

McGinn, A. (2008b, April 23). Off stage, it's all psychological as former finalist pursues degree; Stephanie Edwards. *USA Today.* Retrieved August 1, 2009, from the Lexis Nexis database.

McGinn, A. (2008c, May 8). Everyday job is hard to swallow: Idol Spotlight; Where are they now? Rickey Smith. *USA Today.* Retrieved August 1, 2009, from the Lexis Nexis database.

McLaughlin, M. (2008, February 22). A work ethic worth idolizing: He didn't steal the show, but Constantine Maroulis keeps his eyes on the prize. *The Washington Post.* Retrieved August 1, 2009, from the Lexis Nexis database.

Moran, A. (1998). *Copycat TV: Globalisation, Program Formats, and Cultural Identity.* Luton, Bedfordshire, UK: University of Luton Press.

Morley, D., and Robins, K. (1995). *Spaces of Identity: Global media, electronic landscapes and cultural boundaries.* New York, NY: Routledge.

Murray, S., and Ouellette, L. (Eds.). (2004) *Reality TV: Remaking television culture.* New York, NY: New York University Press.

Newcomb, H. (2007). *Television: The critical view* (7th ed). New York, NY: Oxford University Press.

Nichols, B. (1994). *Blurred boundaries: Questions of meaning in contemporary culture*. Indianapolis, IN: Indiana University Press.

Oldenburg, A. (2002, June 18). Hosts of *American Idol* keep young hopes alive. *USA Today*. Retrieved August 1, 2009, from the Lexis Nexis database.

Parry-Giles, T., and Parry-Giles, S. (2006). *The prime-time presidency: The West Wing and U.S. nationalism.* Chicago, IL: University of Illinois Press.

Pinsker, B. (2007, March 24). 'Idol' worship. *Billboard*, 119 (12), 33–38. Retrieved April 14, 2007, from the Academic Search Premier database.

Price, M. (1995). *Television, the public sphere and national identity*. New York, NY: Oxford University Press.

Pullen, C. (2004). The household, the basement, and *The Real World*: Gay identity in the constructed reality environment. In S. Holmes and D. Jermyn (Eds.), *Understanding reality television* (211–232). New York, NY: Routledge.

Pullen, K. (2000). I-love-Xena.com: Creating online fan communities. In D. Gauntlett, (ed.), *Web.studies: Rewiring media studies for the digital age* (52–61). London, UK: Arnold.

Putnam, R. (2000). *Bowling alone: The collapse and revival of the American community*. New York, NY: Simon and Shuster.

Raphael, C. (2004). The political economic origins of reali-TV. In S. Murray and L. Ouellette (Eds.), *Reality TV: Remaking television culture* (119–36). New York, NY: New York University Press.

Rein, I., Kotler, P., and Stoller, M. (1987). *High visibility*. New York, NY: Dodd, Mead, and Co.

Resources: Who owns what. (2008, March 7). *Columbia Journalism Review*. Retrieved March 7, 2008 from www.cjr.org/resources/.

Riesman, D. (1950). *The lonely crowd.* New Haven, CT: Yale University Press.

Rodman, S. (2008, April 24). 'Idol' style: mixed reviews. *The Boston Globe*. Retrieved August 1, 2009, from the Lexis Nexis database.

Rojek, C. (2001) *Celebrity*. London, UK: Reaktion.

Rosen, C. (2007, March 24). The industry impact. *Billboard*, 119 (12), 42–46. Retrieved April 14, 2007, from the Academic Search Premier database.

Schapiro, R. (2008, February 17). Budding star aims to break the *American Idol* mold. *Daily News (New York)*. Retrieved August 1, 2009, from the Lexis Nexis database.

Schiller, H. (1976). *Communication and cultural domination.* White Plains, NY: International Arts and Sciences Press.

Schiller, H. (1992). *Mass communications and American empire*. Boulder, CO: Westview Press.

Schlesinger, P. (1991). *Media, state, and nation: political violence and collective identities.* London, UK: Sage Publications.

Schneider, M. (2000, July 17). Whiz kid hits global jackpot. *Variety*.

Schneider, M. (2002, June 13). Auds worship 'Idol' for Fox victory. *Daily Variety*. Retrieved August 1, 2009, from the Lexis Nexis database.

Schneider, M. (2008, February 19). 'Idol' harmonizing with Apple via deal. *Daily Variety*. Retrieved August 1, 2009, from the Lexis Nexis database.

Serjeant, J. (2008, December 17). "American Idol" is most time-shifted show. *Reuters UK*. Retrieved from uk.reuters.com/article/televisionNews/idUKTRE4BF7EU20081217? pageNumber=2&virtualBrandChannel=0.

Serpe, G. (2007, May 24). "Idol" ratings not so sparkly. *www.eonline.com*. Retrieved March 5, 2008, from www.eonline.com/news/article/index.jsp?uuid=e1ff5d86–3-e48–4612–9374–5fdba5718ceb.

Serwer, A., and Hajim, C. (2005, December 12). The man who bought Elvis. *Fortune*, 152 (12), 80–96. Retrieved April 14, 2007, from the Academic Search Premier database.

Shales, T. (2002, July 30). *American Idol*: Fool's gold from Fox. *The Washington Post*. Retrieved August 1, 2009, from the Lexis Nexis database.

Sillars, M., and Gronbeck, B. (2001). *Communication criticism: Rhetoric, social codes, cultural studies*. Prospect Heights, IL: Waveland Press.

Silverstone, R. (1988). Television myth and culture. In J. Carey (Ed.), *Media, myths, and narrative* 20–47). Beverly Hills, CA: Sage Publications.

Snyder, P. (Director). (2008a, February 18). *The big story with John Gibson*. [Television Broadcast]. New York, NY: FOX. Transcript retrieved August 1, 2009, from the Lexis Nexis database.

Snyder, P. (Director). (2008b, February 21). *The big story with John Gibson*. [Television Broadcast]. New York, NY: FOX. Transcript retrieved August 1, 2009, from the Lexis Nexis database.

Sony BMG Music Entertainment. (2007). Retrieved April 16, 2007 from www.sonybmg.com.M.

Stahl, M. (2004). A moment like this: *American Idol* and narratives of meritocracy. In C. Washbourne and M. Derno (Eds.), *Bad music: The music we love to hate*. New York: Routledge.

Stakutis, C., and Webster, J. (2005). *Inescapable data: Harnessing the power of convergence*. New York, NY: IBM Press.

Steinberg, B. (2010). From ad age, the annual list of what the broadcast nets are getting from advertisers for each primetime show this season. *Advertising Age*. Retrieved April 19, 2011 from www.tvweek.com/news/2010/10/simon_who_idol_spots_still_pri.php.

Stephens, R. (2004). Socially soothing stories? Gender, race and class in TLC's *A Wedding Story* and *A Baby Story*. In S. Holmes and D. Jermyn (Eds.), *Understanding reality television* (191–210). New York, NY: Routledge.

Storey, J. (2003). Introduction: The study of popular culture and cultural studies. In J. Storey (Ed.), *Cultural theory and popular culture* (2nd ed), (x–xvii). New York, NY: Pearson.

Strinati, D. (1995). *An introduction to theories of popular culture*. New York, NY: Routledge.

Tajfel, H. (1984). (Ed.). *The social dimension: European developments in social psychology*. New York, NY: Cambridge University Press.

Tajfel, H., Jaspars, J., and Fraser, C. (1984). The social dimension in European social psychology. In H. Tajfel (Ed.), *The social dimension: European developments in social psychology* (1–8). New York, NY: Cambridge University Press.

Tajfel, H., and Turner, J. (1986). The social identity theory of intergroup behavior. In S. Worchel and W. G. Austin (Eds.), *Psychology of intergroup relations* (7–24). Chicago, IL: Nelson-Hall Publishers.

The celebrity 100: #22 Simon Cowell. (2008, June 11). *Forbes*. Retrieved August 10, 2009 from www.forbes.com/lists/2008/53/celebrities08_Simon-Cowell_P3QW.html.

The Nielsen Company measures the *American Idol* phenom. (2008, May 18). [Press release]. New York, NY: New York.

Thottam, J. (2006, May 6). *The 2006 Time 100*. Time Magazine.

Tincknell, E., and Raghuram, P. (2004). *Big Brother*: Reconfiguring the "active" audience of cultural studies? In S. Holmes and D. Jermyn (Eds.), *Understanding reality television* (252–69). New York, NY: Routledge.

Tracking advertising dollars: As upfront begins, learn who leads way in spending. (2008, May 5). *Television Week*. Retrieved August 1, 2009, from the Lexis Nexis database.

Traub, J. (2008, March 9). *The celebrity solution*. The New York Times Magazine.

Tuchman, G. (1978). *Making news*. New York, NY: The Free Press.

Tucker, N. (2007, November 1). Reality looms: Writers' strike could change pace of television. *The Washington Post*. Retrieved October 27, 2009, from www.washingtonpost.com/wpdyn/content/article/2007/10/31/AR2007103103054.html.

Turner, G. (2004). *Understanding celebrity*. Thousand Oaks, CA: Sage Publications.

Turner, J. (1996). Henri Tajfel: An introduction. In P. Robinson (Ed.), *Social groups and identities: Developing the legacy of Henri Tajfel* (1–24). Boston, MA: Butterworth-Heinemann.

Turner, J., and Bourhis, R. (1996). Social identity, interdependence, and the social group: A reply to Rabbie et al. In P. Robinson (Ed.), *Social groups and identities: Developing the legacy of Henri Tajfel* (25–64). Boston, MA: Butterworth-Heinemann.

Turner, V. (1981). Social dramas and stories about them. In W.J.T. Mitchell (Ed.), *On narrative* (137–164). Chicago, IL: University of Chicago Press.

Valisno, J. (2008, January 15). "American Idol" debuts at Q-11. *Business World*. Retrieved August 1, 2009, from the Lexis Nexis database.

van Dijk, T. (1991). The interdisciplinary study of news as discourse. In K. Jensen and N. Jankowski (Eds.), *A handbook of qualitative methodologies for mass communication research* (108–120). New York, NY: Routledge.

Waisbord, H. (2007). McTV: Understanding the global popularity of television formats. In H. Newcomb (Ed.), *Television: The critical view* (7th ed), (375–96). New York, NY: Oxford University Press.

Wald, J. (Executive Producer). (2002a, June 27). *Today* [Television Broadcast]. New York, NY: NBC. Transcript retrieved August 1, 2009, from the Lexis Nexis database.

Wald, J. (Executive Producer). (2002b, July 2). *Today* [Television Broadcast]. New York, NY: NBC. Transcript retrieved August 1, 2009, from the Lexis Nexis database.

Wald, J. (Executive Producer). (2002c, August 30). *Today* [Television Broadcast]. New York, NY: NBC. Transcript retrieved August 1, 2009, from the Lexis Nexis database.

Wald, J. (Executive Producer). (2008, February 19). *Today* [Television Broadcast]. New York, NY: NBC. Transcript retrieved August 1, 2009, from the Lexis Nexis database.

Wang, C. (2004, January 12). *Newlywed Bliss. People,* 61 (1). Retrieved April 14, 2007, from the Academic Search Premier database.

Waxman, S. (2002, July 28). Song and dance and derision: Would-be stars risk humiliation on 'American Idol.' For Fox, it's paying off. *The Washington Post*. Retrieved August 1, 2009, from the Lexis Nexis database.

White, H. (1981). The value of narrativity in the representation of reality. In W.J.T. Mitchell (ed.), *On narrative* (1–24). Chicago, IL: University of Chicago Press.

White, H. (1987). *The content of the form: Narrative discourse and historical representation.* Baltimore, MD: Johns Hopkins University Press.

Williams, R. (1981). *Politics and letters: Interviews with New Left Review*. London, UK: Verso.

Williams, R. (1985). *Keywords*. New York, NY: Oxford University Press.

Williamson, J. (1978) *Decoding Advertisements*. London, UK: Marion Boyars.

Wright, W. (1975). *Sixguns and society: A structural study of the Western*. Berkeley, CA: University of California Press.

Wyatt, E. (2008a, March 11). Familiar formula for 'Idol': Popular yet predictable. *The New York Times*. Retrieved August 1, 2009, from the Lexis Nexis database.

Wyatt, E. (2008b, April 7). Where 'Idol's' charitable arm reaches. *The New York Times*. Retrieved August 1, 2009, from the Lexis Nexis database.

Wyatt, E. (2008c, April 22). American idolatry: If only reality were this well organized. *The New York Times*. Retrieved August 1, 2009, from the Lexis Nexis database.

Wyatt, E. (2008d, May 1). 'Idol' judge admits rehearsal guided her. *The New York Times*. Retrieved August 1, 2009, from the Lexis Nexis database.

Yourse, R. (2008, February 20). 'Idol,' IPod overkill. It's official: It's an "American Idol" world. *The Washington Times*. Retrieved August 1, 2009, from the Lexis Nexis database.

Zeleny, J. (2007, December 9). Oprah Winfrey hits campaign trail for Obama. *The New York Times*. Retrieved October 22, 2009, from www.nytimes.com/2007/12/09/us/politics/09oprah.html.

Zelizer, B. (1992). *Covering the body: The Kennedy assassination, the media and the shaping of collective memory*. Chicago, IL: University of Chicago Press.

Zelizer, B. (1997). Has communication explained journalism? In D. Berkowitz (Ed.), *Social meanings of news* (23–30). Thousand Oaks, CA: Sage.

Index

About the Author

Amanda Scheiner McClain is an assistant professor of communications at Holy Family University in Philadelphia, Pennsylvania. She earned a PhD in mass media and communication from Temple University and a BA in communication from the University of Pennsylvania. McClain enjoys working as the television area chair for both the Mid-Atlantic and National Popular/American Culture Association conferences, as well as contributing to many other academic organizations. Her research interests include popular culture, reality TV, social media, collective identity, media narratives, and representations of women.

www.ingramcontent.com/pod-product-compliance
Ingram Content Group UK Ltd.
Pitfield, Milton Keynes, MK11 3LW, UK
UKHW012037150926
475560UK00001B/50

9 780739 164334